Learning from Accidents in I̶   ̶

Learning from Accidents in Industry

Trevor Kletz
DSc, FEng, FRSC, FIChemE

Butterworths
London Boston Singapore Sydney Toronto Wellington

First published 1988

© **Butterworth & Co. (Publishers) Ltd, 1988**

British Library Cataloguing in Publication Data

Kletz, Trevor A.
    Learning from accidents in industry.
    1. Industrial safety
    I. Title
    363.1'17'072   T55

ISBN 0-408-02696-0

Library of Congress Cataloging-in-Publication Data

Kletz, Trevor A.
    Learning from accidents in industry/Trevor A. Kletz.
        p.   cm.
    Bibliography: p.
    Includes index.
    ISBN 0-408-02696-0
    1. Chemical industry—Accidents.   2. Industrial accidents,
3. Industrial accidents—Investigation.   I. Title.
HD7269.C45K43   1988
363.1'165—dc19   87-34155

Photoset by Butterworths Litho Preparation Department
Printed and bound in Great Britain by Anchor Brendon Ltd., Tiptree, Essex

Contents

Acknowledgements

Thanks are due to the companies where the accidents described in this book occurred for permission to describe them, so that we may all learn from them, to the Leverhulme Trust for financial support, to Loughborough University of Technology for giving me the opportunity to develop and record some of the knowledge I acquired during my 38 years in the chemical industry, to Professor F. P. Lees who read the book in manuscript and made many valuable suggestions, and to Mr E. S. Hunt for assistance with Chapter 15.

The book is dedicated to all those killed or injured in the accidents, in the hope that others will learn from their misfortunes.

Forethoughts

It is the success of engineering which holds back the growth of engineering knowledge, and its failures which provide the seeds for its future development.

D. I. Blockley and J. R. Henderson, *Proc. Inst. Civ. Eng.* Part 1, Vol. 68, Nov. 1980, p. 719

What has happened before will happen again. What has been done before will be done again. There is nothing new in the whole world.

Ecclesiastes, 1, 9 (Good News Bible)

What worries me is that I may not have seen the past here – perhaps I have seen the future.

Elie Wiesel

Below, distant, the roaring courtiers
rise to their feet – less shocked than irate.
Salome has dropped the seventh veil
and they've discovered there are eight.

Danny Abse, *Way out in the Centre*

. . . But if so great desire
Moves you to hear the tale of our disasters
Briefly recalled . . .
However I may shudder at the memory
And shrink again in grief, let me begin.

Virgil, *The Aeneid*

Introduction

Accident investigation is like peeling an onion or, if you prefer a more poetic metaphor, the dance of the seven veils. Beneath one layer of causes and recommendations, there are other, less superficial, layers. The outer layers deal with the immediate technical causes while the inner layers are concerned with ways of avoiding the hazards and with the underlying causes such as weaknesses in the management system. Very often only the outer layers are considered and thus we fail to use all the information for which we have paid the high price of an accident. The aim of this book is to show, by analysing accidents that have occurred, mainly but not entirely in the chemical industry, how we can learn more from accidents and thus be better able to prevent them occurring again. The incidents discussed range from the trivial to major accidents such as Flixborough, Seveso and Bhopal. The book should therefore interest all those concerned with the investigation of accidents, of whatever sort, and all those who work in the process industries, whether in design, operations or loss prevention.

I am not suggesting that the immediate causes of an accident are any less important than the underlying causes. All must be considered if we wish to prevent further accidents, as the examples will show. But putting the immediate causes right will prevent only the last accident happening again; attending to the underlying causes may prevent many similar accidents.

Finding the facts

This book is not concerned with the collection of information about accidents but with the further consideration of facts already collected. Those interested in the collection of information should consult a paper by Craven[1] or, if sabotage is suspected, papers by Carson and Mumford[2]. Nevertheless, it may be useful to summarize a few points that are sometimes overlooked[3].

(1) Try not to disturb evidence that may be useful to experts who may be called in later. If equipment has to be moved, for example to make the plant safe, then photograph it first. In the UK a factory inspector may direct that things are left undisturbed 'for so long as is reasonably necessary for the purpose of any examination or investigation'.

(2) Draw up a list of everyone who may be able to help, such as witnesses, workers on other shifts, designers, technical experts etc.

(3) Do not question witnesses in such a way that you put ideas into their minds. Try to avoid questions to which the answer is 'yes' or 'no'. It is easier for witnesses to say 'yes' or 'no' than to enter into prolonged discussions, especially if they are suffering from shock.
(4) Avoid, at this stage (preferably at any stage; see later), any suggestion of blame.
(5) Inform any authorities who have to be notified (in the UK a wide variety of dangerous occurrences have to be notified to the Health and Safety Executive under *The Reporting of Injuries, Diseases and Dangerous Occurrences Regulations (1985)*) and the insurance company, if claims are expected.
(6) Record information, quantitative if possible, on damage and injuries so that others can use it for prediction.

Avoid the word 'cause'

Although I have used this word it is one I shall use sparingly when analysing accidents, for three reasons.

(1) If we talk about causes we may be tempted to list those we can do little or nothing about. For example, a source of ignition is often said to be the cause of a fire. But when flammable vapour and air are mixed in the flammable range, experience shows that a source of ignition is liable to turn up, even though we have done everything possible to remove known sources of ignition (see Chapter 4). The only really effective way of preventing an ignition is to prevent leaks of flammable vapour. Instead of asking, 'What is the cause of this fire?', we should ask 'What is the most effective way of preventing another similar fire?'. We may then think of ways of preventing leaks.

Another example: Human error is often quoted as the cause of an accident but as I try to show in my book *An Engineer's View of Human Error*[4], there is little we can do to prevent people making mistakes, especially those due to a moment's forgetfulness. If we ask 'What is the cause of this accident?', we may be tempted to say 'human error' but if we ask 'What should we do differently to prevent another accident?' we are led to think of changes in design or methods of operation.
(2) The word 'cause' has an air of finality about it that discourages further investigation. If a pipe fails, for example, and the cause is said to be corrosion, we are tempted to think that we know why it failed. But to say that a pipe failure was due to corrosion is rather like saying that a fall was due to gravity. It may be true but it does not help us to prevent further failures. We need to know the answers to many more questions. Was the material of construction specified correctly? Was the specified material actually used? Were operating conditions the same as those assumed by the designers? What corrosion monitoring did they ask for? Was it carried out? Were the results ignored? And so on.
(3) The word 'cause' implies blame and people become defensive. So instead of saying that an accident was caused by poor design (or maintenance or operating methods) let us say that it could be

prevented by better design (or maintenance or operating methods). We are reluctant to admit that we did something badly but we are usually willing to admit that we could do it better.

However, the main point I wish to make is that whether we talk about causes or methods of prevention, we should look below the immediate technical changes needed, at the more fundamental changes such as ways of avoiding the hazard and ways of improving the management system.

The irrelevance of blame

If accident investigations are conducted with the objective of finding culprits and punishing them, then people do not report all the facts, and who can blame them? We never find out what really happened and are unable to prevent it happening again. If we want to know what happened we have to make it clear that the objective of the inquiry is to establish the facts and make recommendations and that nobody will be punished for errors of judgement or for forgetfulness, only for reckless indifference to the safety of others. Occasional negligence may go unpunished, but this is a small price to pay to prevent further accidents. An accident may show that someone does not have the ability to carry out a particular job and he may have to be moved, but this is not punishment and should not be made to look like punishment.

In fact very few accidents are the result of negligence. Most human errors are the result of a moment's forgetfulness or aberration, the sort of error we all make from time to time. Others are the result of errors of judgement, inadequate training or instruction or inadequate supervision[4].

Accidents are rarely the fault of a single person. Responsibility is usually spread amongst many people. To quote from an official UK report on safety legislation[5]:

'The fact is – and we believe this to be widely recognized – the traditional concepts of the criminal law are not readily applicable to the majority of infringements which arise under this type of legislation. Relatively few offences are clear cut, few arise from reckless indifference to the possibility of causing injury, few can be laid without qualification at the door of a single individual. The typical infringement or combination of infringements arises rather through carelessness, oversight, lack of knowledge or means, inadequate supervision, or sheer inefficiency. In such circumstances the process of prosecution and punishment by the criminal courts is largely an irrelevancy. The real need is for a constructive means of ensuring that practical improvements are made and preventative measures adopted.'

How can we encourage people to look for underlying causes?

First they must be convinced that the underlying causes are there and that it will be helpful to uncover them. Reading this book may help. A better way is by discussion of accidents that have occurred and the action needed to prevent them happening again. The discussion leader describes an accident very briefly; those present question him to establish the rest of the facts and then say what *they think* ought to be done to prevent it happening again. The UK Institution of Chemical Engineers provide sets of notes and slides for use in such discussions[6]. The incidents in this book may also be used. It is better, however, to use incidents which have occurred in the plant in which those present normally work. Some discussion groups concentrate on the immediate causes of the incidents discussed; the discussion leader should encourage them to look also at the wider issues.

When fresh accidents have to be investigated it is helpful to have in the investigating team at least one person from another part of the organization. Such a person is much more likely than those closely involved to see the wider issues and the relevance of the incident to other plants.

After a time, it becomes second nature for people who have looked for the less obvious ways of preventing accidents, either in discussion or in real situations, to continue to do so without prompting.

Most of the recommendations described in this book were made during the original investigation but others only came to light when the accidents were later selected for discussion in the way I have described.

In the book the presentations differ a little from chapter to chapter, to avoid monotony and to suit the varying complexity of the accounts. Thus in discussing fires and explosions, a discussion of the source of ignition may be followed by recommendations for eliminating it. In other cases, all the facts are described first and are followed by all the recommendations.

Occasionally questions are asked to which there are no clear or obvious answers.

Is it helpful to use an accident model?

Many people believe that it is and a number of models have been described. For example, according to Houston[7,8] three input factors are necessary for an accident to occur; target, driving force and trigger. For example, consider a vessel damaged by pressurization with compressed air at a pressure above the design pressure (as in the incident described in Chapter 7). The driving force is compressed air, the target a vessel to which it is connected and the trigger the opening of the connecting valve. The development of the accident is determined by a number of parameters: the contact probability (the probability that all the necessary input factors are present), the contact efficiency (the fraction of the driving force which reaches the target) and the contact time. The model indicates a number of ways in which the probability or severity of the accident may be reduced. One of the input factors may be removed or the effects of the parameters minimized. Other models have been described by Pope[9] and Ramsey[10].

Personally I have not found such models useful. I find that time may be spent struggling to fit the data into the framework and that this distracts from the free-ranging thinking required to uncover the less obvious ways of preventing the accident. A brainstorming approach is needed. I do give in the Appendix to Chapter 17 a list of questions that may help some people to look below the surface but they are in no sense a model. Use models by all means if you find them useful but do not become a slave to them. Disregard them if you find that they are not helping you.

However, although I do not find a general model useful, I do find it helpful to list the chain of events leading up to an accident, and these chains are shown for each accident that is discussed in detail. They show clearly that the chain could have been broken, and the accident prevented, at any point. At one link in the chain the senior managers of the company might have prevented the accident by changing their organization or philosophy; at another link the operator or craftsman might have prevented it by last-minute action; designers, managers and foremen also had their opportunities. The chains remind us that we should not use inaction by those above (or below) us as an excuse for inaction on our part. The explosion described in Chapter 4 would not have occurred if the senior managers had been less insular. Equally it would not have occurred if a craftsman had made a joint with greater skill.

The chain diagrams illustrate the onion effect by different typefaces. Attention to the underlying causes may break the chain at various points, not just at the beginning, as the diagrams will show.

There are no right answers

If the incidents described in this book are used as subjects for discussion, as described earlier, it must be emphasized that there are no right answers for the group to arrive at. The group may think that my recommendations go too far, or not far enough, and they may be right. How far we should go is a matter of opinion. What is the right action in one company may not be right for another which has a different culture or different working practices. I have not tried to put across a set of answers for specific problems, a code or a standard method for investigating accidents, but rather a way of looking at them. I have tried to preserve the divergence of view which is typical of the discussions at many inquiries so that the book has something of an oral character.

Nevertheless, while the primary purpose of the book is to encourage people to investigate accidents more deeply, I hope that the specific technical information given in the various chapters will also be useful, in helping readers deal with similar problems on their own plants. The incidents discussed did not have exotic causes – few have. After most of them people said, 'We ought to have thought of that before'.

Prevention first

The investigations described in this book should ideally have been carried out when the plants were being designed so that modifications, to plant design or working methods, could have been made before the accidents occurred, rather than after. Samuel Coleridge described history as a lantern on the stern, illuminating the hazards the ship has passed through rather than those that lie ahead. It is better to see the hazards afterwards than not see them at all, as we may pass the same way again, but it is better still to see them when they still lie ahead. There are methods available helping us to foresee hazards but they are beyond the scope of this book.

Briefly, those that I consider most valuable are:

- Hazard and operability studies (hazops)[11, 12] at the flowsheet and detailed design stages.
- A variation of the technique at the earlier, conceptual design stage[13] (see Chapter 17).
- Detailed inspection during and after construction to make sure that the design has been followed and that details not specified in the design have been constructed in accordance with good engineering practice[14] (see Chapter 16).
- Safety audits on the operating plant[15].

Record all the facts

Investigating teams should place on record all the information they collect and not just that which they use in making their recommendations. Readers with a different background, experience or interests may then be able to draw additional conclusions from the evidence. Chapter 14 gives some examples. UK official reports are usually outstanding in this respect. The evidence collected is clearly displayed, and then conclusions are drawn and recommendations made. Readers may draw their own conclusions, if they wish to do so. In practice they rarely draw contradictory conclusions but they may draw additional and deeper ones.

It is usual in scholarly publications to draw all the conclusions possible from the facts. Compare, for example, the way archaeologists draw pages of deductions from a few pottery fragments. In this respect most writing on accidents had not been scholarly, authors often being content to draw only the most obvious messages.

Nevertheless, reports should not be too verbose or busy people will not read them. The ideal is two reports: one giving the full story and the other summarizing the events and drawing attention to those recommendations of general applicability which apply outside the unit where the incident occurred. (See the Appendix to Chapter 15).

Other information to include in accident reports

Information which it is useful to include in accident reports, but is often not included, is:

- **Who** is responsible for carrying out the recommendations?
- **When** will they be complete? The report can then be brought forward at this time.
- **How much** will they cost, in money and other resources (for example, two design engineers for three weeks or one electrician for three days)? We can then see if the resources are likely to be available. In addition, though safety is important, we should not write blank cheques after an accident. If the changes proposed are expensive we should ask if the risk justifies the expenditure or if there is a cheaper way of preventing a recurrence. The law in the UK does not ask us to do everything possible to prevent an accident, only what is 'reasonably practicable'.

Many people feel that an accident report is incomplete if it does not recommend a change to the plant, but sometimes altering the hardware will not make another accident less likely. If protective equipment has been neglected, will it help to install more protective equipment? (See Chapter 6).

Acknowledgements

I would like to thank the many companies where the incidents I have described occurred for permitting me to describe their mistakes, and the very many colleagues with whom I have discussed these accidents and who suggested the various comments, some emphasizing the immediate causes and others the underlying ones. All are valuable. My colleagues – particularly those who attended my discussions – are the real authors of this book. I am merely the amanuensis. Rabbi Judah the Prince (ca 135–217 AD) said, 'Much have I learnt from my teachers, more from my colleagues and most of all from my students'.

I do not name the products made on the plants where the incidents occurred, partly to preserve their anonymity but also for another reason: if I said that an explosion occurred on a plant manufacturing acetone, readers who do not use acetone might be tempted to ignore that report. In fact, most of the recommendations apply to most plants, regardless of the materials they handle. To misquote the well-known words of the poet John Donne,

No plant is an Island, entire of itself; every plant is a piece of the Continent, a part of the main. Any plant's loss diminishes us, because we are involved in the Industry; and therefore never send to know for whom the inquiry sitteth; it sitteth for thee.

Descriptions of most of the accidents described in this book have appeared before but scattered throughout various publications, often in a different form. References are given at the end of each chapter and thanks are due to the original publishers for permission to quote from them.

References

1. CRAVEN, A. D. 'Fire and explosion investigations on chemical plants and oil refineries', in *Safety and Accident Investigations in Chemical Operations,* edited by H. H. Fawcett and W. S. Wood, 2nd edition, Wiley, New York, 1982, p. 659.
2. CARSON, P. A., MUMFORD, C. J. and WARD, R. B. *Loss Prevention Bulletin,* No. 065, Oct. 1985, p. 1 and No. 070, August 1986, p. 15.
3. FARMER, D. *Health and Safety at Work,* Vol. 8, No. 11, Nov. 1986, p. 54.
4. KLETZ, T. A. *An Engineer's View of Human Error,* Institution of Chemical Engineers, Rugby, UK, 1985.
5. *Safety and Health at Work: Report of the Committee 1970–1972* (The Robens Report), Her Majesty's Stationery Office, London, 1972, paragraph 261.
6. *Hazard Workshop Modules,* Nos. 001–008, Institution of Chemical Engineers, Rugby, UK, 1978–1986. They deal with: Over- and under-pressuring of vessels; Plant modifications; Fires and explosions; Preparation for maintenance (2 sets); Furnace fires and explosions; Handling emergencies; Human error.

7. HOUSTON, D. E. L. *Major Loss Prevention in the Process Industries,* Institution of Chemical Engineers, Rugby, UK, 1971, p. 210.

8. LEES, F. P. *Loss Prevention in the Process Industries,* Butterworths, London, 1980, Vol. 1, Section 2.1; Vol. 2, Section 27.4.

9. POPE, W. C. 'In case of accident, call the computer', in *Selected Readings in Safety,* edited by J. T. Winder, Academy Press, Macon, Georgia, 1973, p. 295.

10. RAMSEY, J. D. 'Identification of contributory factors in occupational injury and illness', as ref. 9, p. 328.

11. KLETZ, T. A. *Hazop and Hazan – Notes on the Identification and Assessment of Hazards,* Institution of Chemical Engineers, Rugby, UK, 2nd edition, 1986.

12. As ref. 8, Vol. 1, Section 8.9.

13. KLETZ, T. A. *Cheaper, Safer Plants,* Institution of Chemical Engineers, Rugby, UK, 2nd edition, 1985.

14. KLETZ, T. A. *Plant/Operations Progress,* Vol. 3, No. 1, Jan. 1984, p. 19.

15. As ref. 8, Vol. 1, Section 8.1.

Chapter 1

Two simple incidents

This chapter analyses two simple accidents in order to illustrate the methods of 'layered' accident investigation. They also show that we should investigate all accidents, including those that do not result in serious injury or damage, as valuable lessons can be learned from them. 'Near misses', as they are often called, are warnings of coming events. We ignore them at our peril, as next time the incidents occur the consequences may be more serious. Engineers who brush aside a small fire as of no consequence are like the girl who said by way of excuse that it was only a small baby. Small fires, like small babies, grow into bigger ones.

A small fire

A pump had to be removed for repair. The bolts holding it to the connecting pipework were seized and it was decided to burn them off. As the plant handled flammable liquids, the pump was surrounded by temporary sheets of a flame-resistant material and a drain about a metre away was covered with a polyethylene sheet. Sparks burned a hole in it and set fire to the drain. The fire was soon extinguished and no-one was hurt. The atmosphere in the drain had been tested with a flammable gas detector two hours before burning started but no gas was detected, probably because flammable gas detectors will work only when oxygen is present, and there was too little oxygen below the sheet. It is possible, however, that conditions changed and flammable vapour appeared in the drain during the two hours that elapsed before burning started.

First layer recommendations: Preventing the accident

In future:
- Cover drains with metal or other flame-resistant sheets before allowing welding or burning nearby.
- Test the atmosphere *above* the sheets, not below them.
- Test the atmosphere immediately before welding starts, not several hours before. In addition, install a portable flammable gas detector which will sound an alarm if conditions change and gas appears while welding or burning are in progress.

These recommendations apply widely, not just on the unit where the fire occurred, so the information should be passed on to other plants.

Second layer recommendations: Avoiding the hazard

Why were the bolts seized? Lubricants which prevent seizure, even at the high temperatures used in this case, are available. Whose job is it to see the need for such lubricants, and that they are used?

In an area where flammable liquids or gases are handled, seized bolts would normally be cut off rather than burned off. In the present case access was so poor that it was decided to burn them off. Why was access so poor? The normal policy in the company was to build a model of the plant before detailed design was carried out and to review access for maintenance on the model (as well as access for operations, means of escape and many other matters). What went wrong in this case? Was the model review really carried out and were operating and maintenance people present?

Third layer recommendations: Improving the management system

Did the men on the job understand that flammable gas detectors will not detect flammable gas unless it is mixed with air (or oxygen) in the flammable range? Many operators do not understand this limitation of flammable gas detectors. Is this point covered in their training? What is the best way of putting it across so that people will understand and remember?

The plant instructions said that drains must be covered with flame-resistant sheets when welding or burning takes place nearby. Over the years everyone had got into the habit of using polyethylene sheets. Did the managers not notice? Or did they notice and turn a blind eye? ('I've got more important things to do than worry about the use of the wrong sort of sheet.') To prevent the fire, it needed only one manager to keep his eyes open, see that polyethylene sheets were being used and ask why. On this plant do the managers spend a few hours per day out on the site with their eyes open or do they feel that wandering round the site can be left to the foremen and that their job is to sit in their office thinking about technical problems?

Note that I am using the word 'manager' in the UK sense of anyone working at the professionally qualified level, and that it includes people who in many US companies would be called supervisors or superintendents.

Some readers may feel that I am making heavy weather of a minor incident but questions such as these are unlikely to be asked unless an incident or series of incidents throw them into focus. Obviously the answers given and the changes made will depend on whether the incident is an isolated one, or whether other incidents have also drawn attention to weaknesses in training, managerial powers of observation etc.

The investigating team for an incident such as this would not normally contain any senior managers and we would not expect the unit manager or supervisor to think of all the second and third layer recommendations. But more senior managers should think of them when they read the report. Nor does it take any longer to think of the deeper recommendations as well as the obvious ones. The resource needed is a realization that such recommendations are possible and necessary, rather than additional time to spend on investigations.

Figure 1.1 summarizes, on a time scale, the events leading up to the accident and the recommendations made. It should be read from the bottom up. First, second and third layer recommendations are indicated by different typefaces. First layer recommendations, the immediate technical ones, are printed in ordinary type, second layer recommendations, ways of avoiding the hazard, are printed in italics, and third layer recommendations, ways of improving the management system, are printed in bold type. The same scheme is followed in later chapters though the allocation

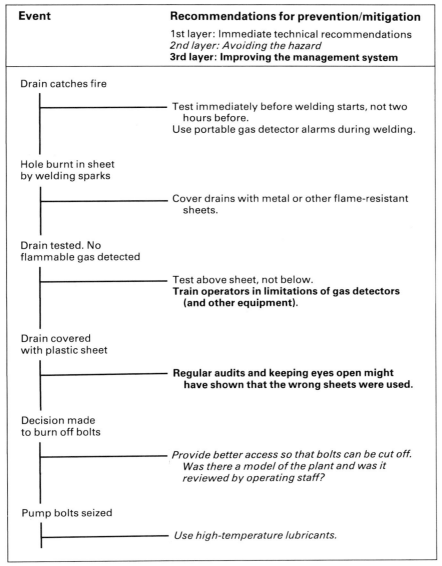

Figure 1.1   Summary of Chapter 1 – first incident

between categories is inevitably in some cases a matter of opinion. Thus hazop is shown as a means of avoiding the hazard but might equally well be considered a means of improving the management system.

The diagram shows us that there were many opportunities of preventing the accident, by breaking the chain of events that lead up to it at different points. Some of the actions had to be taken shortly before the accident occurred, others a long time before. Some of these actions would have removed the immediate causes while others would have removed the hazard or dealt with the weaknesses in the management system which were the underlying causes.

In general, the most effective actions are those at the bottom of the diagram. If we are constructing defences in depth we should make sure that the outer defences are sound as well as the inner ones. Protective measures should come at the bottom of the accident chain and not just at the top. In many of the accidents described later there was too much dependence on the inner defences, the protective measures at the top of the accident chain. When these defences failed there was nothing in reserve.

A mechanical accident

This section describes an accident to a mixer – but it is really about all accidents, so please read it even if you never have to design or operate a mixer.

A mixing vessel of $1\,m^3$ (264 US gallons) capacity was fitted with a hinged, counterweighted lid (Figure 1.2). To empty the vessel the lid was opened (Figure 1.3), the vessel rotated anticlockwise and the contents shovelled out (Figure 1.4). One day the lid fell off and hit the man who was emptying the vessel. Fortunately his injuries were not serious.

It was found that the welds between the lid and its hinges had cracked. It was a fatigue failure, caused by the strains set up by repeated opening and closing of the lid. There was nothing wrong with the original design but the lid had been modified about 10 years before the incident occurred and, in addition, some repairs carried out a few years before had not been to a high enough standard.

Detailed recommendations were made for the repair of the lid. Though necessary, they do not go far enough. If we look at the inner layers of the onion four more recommendations are seen to be necessary.

(1) What is the system for the control of modifications? Is anyone who things he can improve a piece of equipment allowed to do so? Before

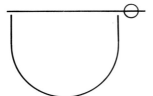

Figure 1.2   The mixing vessel in use

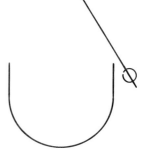

Figure 1.3   The lid is opened

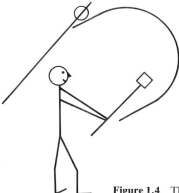

Figure 1.4   The vessel is rotated so that the contents can be removed

any equipment is modified the change should be approved by a professionally qualified engineer, who tries to make sure that the change is to the same standard as the original design and that there are no unforeseen side-effects. This is one of the lessons of Flixborough (Chapter 8). Many other accidents have occurred because plants or processes were modified and no one foresaw the consequences of the change[1,2].

After a modification has been made, the engineer who approved it should inspect the completed work to make sure that his intentions have been followed and that the modification looks right. What does not look right is usually not right and should at least be checked (Figure 1.5).

(2) Why were the repairs not carried out to a high enough standard? Who is (or should be made) responsible for specifying the standard of

When the plant has been modified, have a look at it to make sure that:
- your intentions have been followed
- the modification "looks right"

Figure 1.5

repairs and modifications and checking that work has been carried out to this standard? Does anyone know the original design standard?

(3) Cracks would have been present in the welds for some time before they failed completely and could have been detected if the lid had been inspected regularly. The company concerned registered and inspected all pressure vessels and, under a separate scheme, all lifting gear. However, the mixer was not registered under either scheme, as it operated at atmospheric pressure and so was not a pressure vessel, and it was not recognized as lifting gear. Yet its failure could be as dangerous as the failure of vessels or lifting gear. It should be registered under one of the schemes. It does not matter which, provided the points to be looked for during inspection are noted.

Many other accidents have occurred because equipment was not recognized as coming into one of the categories that should be registered and inspected or treated in some special way. Chapter 7 discusses an accident that occurred because the size of an open vent was reduced without checking that the smaller size would be adequate. No one realized that the vent was the vessel's relief valve and should be treated like a relief valve: its size should not be changed unless we have gone through the same procedure as we would go through before changing the size of a relief valve.

Similarly, if a relief valve has been sized on the assumption that a non-return (check) valve (or two in series) will operate, the non-return valve should be included in the register of relief valves and inspected regularly, say once per year. If a relief valve has been sized on the assumption that a control valve trim is a certain size, this control valve should be included in the relief valve register, its size should not be changed without checking that the new size will be adequate and the valve should be scheduled for regular examination, say once per year, to check that the original trim is still in position. The control valve register should be marked to show that this valve is special.

(4) People should not normally be expected to work underneath heavy suspended objects. This was apparently not known to those who designed, ordered and operated the mixer, though as far back as 1891 the House of Lords (in Smith v Baker & Sons) ruled that it was an unsafe system of work to permit a crane to swing heavy stones over the heads of people working below[3]. The company carried out regular safety audits but though the mixer had been in use for 10 years no-one recognized the hazard. What could be done to improve the audits? Perhaps if outsiders had been included in the audit teams they would have picked up the hazard.

Just as people should not work below equipment which is liable to fall, so they should not work above equipment which is liable to move upwards. At an aircraft factory a man was working above a fighter plane which was nearly complete. The ejector seat went off and the man was killed. In general, potential energy and trapped mechanical energy are as dangerous as trapped pressure and should be treated with the same respect. Before working on a forklift truck or any other mechanical handling equipment we should make sure that it is in the lowest energy state, i.e. in a position in which it is least likely to move as it is being dismantled. If equipment contains springs, they should be released from compression (or extension) before the equipment is dismantled.

These facts show that thorough consideration of a simple accident can set in motion a train of thought that can lead to a fresh look at the way a host of operations are carried out.

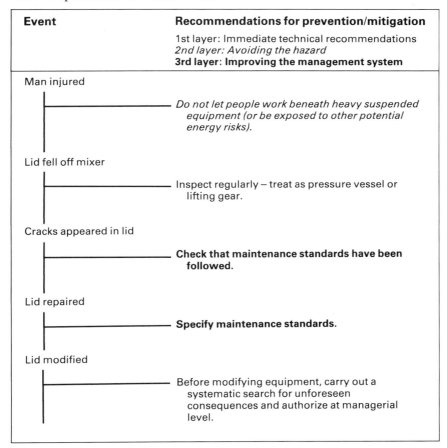

Figure 1.6   Summary of Chapter 1 – second incident

References

The incidents described in this Chapter originally appeared, in a much shorter form, in *Health and Safety at Work,* Vol. 7, No. 1, Jan. 1985, p. 8, and *Occupational Safety and Health,* Vol. 15, No. 2, Feb. 1985, p. 25.

1. KLETZ, T. A. *What went wrong? Case Histories of Process Plant Disasters,* Gulf Publishing Co., 1985, Chapter 2.
2. LEES, F. P. *Loss Prevention in the Process Industries,* Butterworths, 1980, Chapter 21.
3. FARMER, D. *Health and Safety at Work,* Vol. 8, No. 11, Nov. 1986, p. 61.

Protective system failure

To meet a demand from some customers for a product containing less water, a small drying unit was added to a plant which manufactured an organic solvent. The solvent, which was miscible with water, was passed over a drying agent for about eight hours; the solvent was then blown out of the drier with nitrogen and the drier regenerated. There were two driers, one working, one regenerating (Figure 2.1).

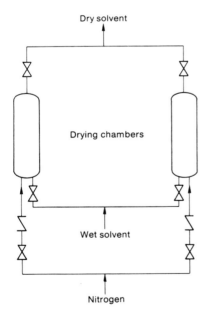

Dry solvent

Drying chambers

Wet solvent

Nitrogen

Figure 2.1 Drying unit in which the accident occurred. (Regeneration lines not shown)

As the drying unit was some distance from the control room the instruments associated with it were mounted on the outdoor control panel shown in Figure 2.2. The top half of the panel contained pneumatic instruments, the lower half electrical equipment associated with the changeover of the driers. The control panel was located in a Zone (Division) 2 area, i.e. an area in which a flammable mixture is not likely to occur in normal operation and, if it does occur, will exist for only a short time (say for a total of not more than 10 hours per year). The electrical equipment could not, at the time of construction, be obtained in a flameproof or non-sparking form suitable for use in a Zone 2 area. It was therefore mounted in a metal cabinet, made from thin metal sheet, which was continuously purged with nitrogen. The nitrogen was intended to keep out any solvent vapour that might leak from the drying unit or the main plant. Such leaks were unlikely, and if they did occur would probably be short-lived, but the Zone 2 classification showed that they could not be ruled out. A pressure switch isolated the electricity supply if the pressure in the cabinet fell below a preset value, originally ½ inch water gauge.

No solvent or other process materials were connected to the control panel.

Despite these precautions an explosion occurred during the commissioning of the drying unit. It had been shut down for a few days and was ready to restart. A young graduate had been given the job of commissioning the unit as his first industrial experience. Standing in the position shown in Figure 2.2 he personally switched on the electricity supply. There was an explosion and the front cover was blown off the metal cabinet, hitting him in the legs. Fortunately no bones were broken and he returned to work after a few days.

For an explosion we need fuel, air (or oxygen) and a source of ignition, and we shall consider these separately before looking at the underlying factors.

Figure 2.2   Instruments controlling the drying unit were located in this outdoor panel. Electrical equipment was purged with nitrogen

The fuel

There was no leak from the drying unit or the main plant at the time and there was no flammable vapour present in the atmosphere. The fuel did not leak into the metal cabinet from outside, the route which had been foreseen and against which precautions had been taken, but entered with the nitrogen. The nitrogen supply was permanently connected to the driers by single isolation valves and non-return (check) valves as shown in Figure 2.1. The gauge pressure of the nitrogen was nominally $40 \, lbf/in^2$ (almost 3 bar) but fell when the demand was high. The gauge pressure in the driers was about $30 \, lbf/in^2$ (2 bar). Solvent therefore entered the nitrogen lines through leaking valves and found its way into the inside of the cabinet. The solvent had to pass through a non-return (check) valve but these valves are intended to prevent gross back-flow, not small leaks. In the photograph of the inside of the cabinet (Figure 2.3), taken immediately after the explosion, the damaged paintwork shows that solvent must have been present for some time. However, solvent vapour and nitrogen will not explode and the solvent alone could not produce an explosive atmosphere.

The air

Air diffused into the cabinet as the nitrogen pressure had fallen to zero for some hours immediately before the accident. The unit was at the end of the nitrogen distribution network and suffered more than most units from deficiencies in the supply. It is difficult to get airtight joints in a cabinet made from thin metal sheets bolted together, and air diffused in through the joints. The solvent may have affected the gaskets in the joints and made them more porous.

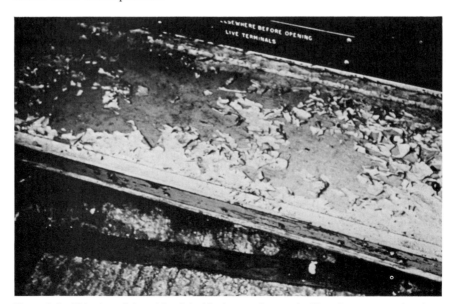

Figure 2.3   This view of the cabinet's inside shows paint attacked by solvent, suggesting that vapour had been getting in for some time

The source of ignition

The source of ignition was clearly electrical, as the explosion occurred when the electricity was switched on. However, the low-pressure switch should have isolated the supply. The reason it did not do so is shown by Figure 2.4, a photograph of the pressure switch with the cover removed. It

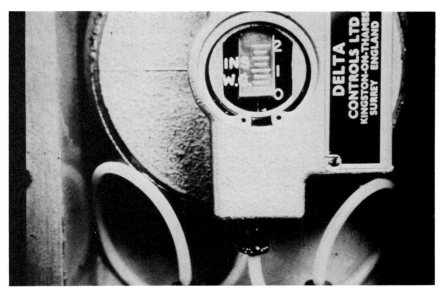

Figure 2.4 Note that on the switch, as shown here with the cover removed, the set-point has been reduced enough to disarm the protective equipment

will be seen that the set-point has been reduced from ½ inch water gauge to zero. The switch cannot operate unless the pressure in the cabinet becomes zero, an impossible situation. The protective equipment has been effectively disarmed (i.e. made inoperable).

The switch was normally covered by a metal cover and the set-point was not visible. Only electricians were authorized to remove the cover.

First layer recommendations

The following recommendations were made during the inquiry immediately following the incident.

The fuel To prevent contamination of the nitrogen it should not be permanently connected to the driers by single valves but by flexes which are disconnected when not in use or by double block and bleed valves. In addition, in case the nitrogen pressure falls while the nitrogen is in use, there should be a low-pressure alarm on the nitrogen supply set a little above the pressure in the driers.

The first recommendation applies whenever service lines have to be connected to process equipment and the second one applies whenever the pressure in a service line is liable to fall below the process pressure. (If the process pressure is liable to rise above the service pressure, there should be a high-pressure alarm on the process line.) Neglect of these precautions has resulted in nitrogen leaks catching fire, air lines setting solid and steam lines freezing.

The accident at Three Mile Island (Chapter 11) was also initiated by back-flow into a service line.

In the longer term a more reliable nitrogen supply should be provided, either by improving the supply to the whole plant or by providing an independent supply to equipment which is dependent on nitrogen for its safety.

The air   It is impossible to make an airtight box from thin metal sheets bolted together. If the nitrogen supply could not be guaranteed then the metal cabinet should have been made more substantial.

The source of ignition   Alterations in the set-points of trips (and alarms) should be made only after authorization in writing at managerial level. They should be recorded and made known to the operators.

Set-points should be visible to the operators; the pressure switch should therefore have a glass or plastic cover. Unfortunately, carrying out this recommendation is not as easy as it sounds. The switch was a flameproof one and could not be modified without invalidating its certification. Redesign had to be discussed with and agreed by the manufacturer and followed by recertification.

All trips (and alarms) should be tested regularly. This was the practice on the plants concerned but as the drying unit was new it had not yet been added to the test schedules. Obviously new equipment (of any sort) should be scheduled for whatever testing and inspection is considered necessary as soon as it is brought into use.

These recommendations also apply to all plants.

Second layer recommendations

After the dust had settled and those concerned had had time to reflect, they asked why the trip had been disarmed. It seemed that the operators had had difficulty maintaining a pressure of ½ inch water gauge in the leaking cabinet. The trip kept operating and shutting down the drying unit. They complained to the electrical department, who reduced the set-point to ¼ inch water gauge. This did not cure the problem. Finally one electrician solved the problem by reducing the set-point to zero. He did not tell anyone what he had done and the operators decided he was a good electrician who had succeeded where the others had failed. After the explosion he chose anonymity.

The designers had not realized how difficult it is to maintain even a slight pressure in a cabinet of thin metal sheets. If they had done so they might have installed a low-flow alarm instead of a low-pressure alarm. In addition they did not know that the nitrogen supply was so unreliable. The plant data sheets showed that a nitrogen supply at a gauge pressure of $40\,lbf/in^2$ was available and they took the data sheets at their word. If a hazard and operability study had been carried out on the design with the unit manager present, then this would probably have come to light. A hazard and operability study was carried out but only on the process lines, not on the service lines. Many other incidents have shown that it is necessary to study service lines as well as process lines[1].

Third layer recommendations

Further recommendations were made when the explosion was selected for discussion by groups of managers and designers, as described in the Introduction. (Some of these recommendations deal with ways of avoiding the hazard and have therefore been classified as second layer in Figure 2.5.)

The cabinet could be pressurized with air instead of nitrogen. The purpose of the nitrogen was to prevent solvent vapour diffusing in from outside. Air could do this equally well and the reliability of the compressed air supply was much better than that of the nitrogen supply. Compressed air was also much cheaper.

Compressed air has another advantage. If anyone had put his head in the cabinet to inspect or maintain the contents, without first making sure that the nitrogen supply was disconnected, he could have been asphyxiated. Compressed air would have removed this hazard. Nitrogen is widely used to prevent fires and explosions but many people have been killed or overcome by it. It is the most dangerous gas in common use[2].

The safety equipment, installed to guard against a rather unlikely hazard – a leak of solvent near the control panel – actually produced a greater hazard. Is there, then, another way of guarding against the original hazard? One possibility is installing flammable gas detectors to detect a leak of solvent and then using the signal from them to isolate the electricity supply.

Did the control panel have to be in a Division 2 area? It did not. It was put in what was believed to be a convenient location without considering the electrical classification of the area. The electrical design engineer then asked for nitrogen purging so that he could meet the requirements of the classification. He did not ask if the control panel could be moved. That was not his job. His job was to supply equipment suitable for the agreed classification. It was no-one's job to ask if it would be possible to change the classification by moving the equipment. In fact, if the control panel had been moved a few metres it would have been in a safe area.

This illustrates our usual approach to safety. When we recognize a hazard – in this case a leak of solvent vapour ignited by electrical equipment – we *add on* protective equipment to control it. We rarely ask if the hazard can be removed.

The man who was injured was, as already stated, a young inexperienced graduate. A more experienced person might have foreseen the hazards and taken extra precautions. Thus the nitrogen pressure should have been checked before the electricity was switched on. It is bad practice to assume that a trip will always work and rely on it. Letting people learn by doing a job and making mistakes is excellent training for them (though not always good for the job) but is not a suitable procedure when hazardous materials

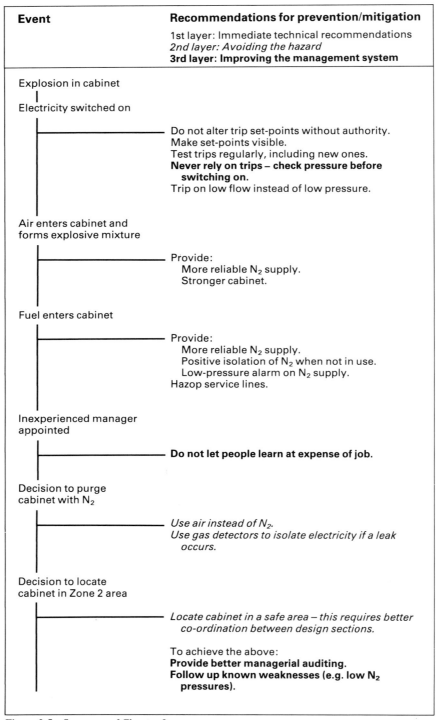

Event	Recommendations for prevention/mitigation
	1st layer: Immediate technical recommendations *2nd layer: Avoiding the hazard* **3rd layer: Improving the management system**

Explosion in cabinet
|
Electricity switched on

Do not alter trip set-points without authority.
Make set-points visible.
Test trips regularly, including new ones.
Never rely on trips – check pressure before switching on.
Trip on low flow instead of low pressure.

Air enters cabinet and forms explosive mixture

Provide:
More reliable N_2 supply.
Stronger cabinet.

Fuel enters cabinet

Provide:
More reliable N_2 supply.
Positive isolation of N_2 when not in use.
Low-pressure alarm on N_2 supply.
Hazop service lines.

Inexperienced manager appointed

Do not let people learn at expense of job.

Decision to purge cabinet with N_2

Use air instead of N_2.
Use gas detectors to isolate electricity if a leak occurs.

Decision to locate cabinet in Zone 2 area

Locate cabinet in a safe area – this requires better co-ordination between design sections.

To achieve the above:
Provide better managerial auditing.
Follow up known weaknesses (e.g. low N_2 pressures).

Figure 2.5   Summary of Chapter 2

are handled. If the young engineer had received some training in loss prevention during his university course, or in a company training scheme, he might have been better prepared for the task he was given. Today, in the UK, though not in most other countries, all undergraduate chemical engineers get some training in loss prevention[3].

Other points

At first sight this explosion seemed to result from the coincidence of four events, three of them unlikely:

(1) A low nitrogen pressure, which allowed solvent to contaminate the nitrogen, followed by
(2) a complete nitrogen failure, which allowed air to diffuse into the cabinet;
(3) disarming of the trip which should have isolated the electricity when the pressure in the cabinet was low and
(4) the triggering event, someone switching on the electricity.

In fact the first three were not events occurring at a point in time but unrevealed faults that existed for long periods; the first had existed on and off for weeks, the second for several hours, from time to time, and the third for days or weeks. It was therefore almost inevitable that sooner or later the triggering event – a true event – would coincide with the three continuing states and an explosion would result.

Accidents are often said to be due to unlikely coincidences, thus implying that people could not reasonably have been expected to foresee them and take precautions. When the facts are established, however, it is usually found that all but one of the events were unrevealed faults that had existed for some time. When the final event occurred, the accident was inevitable.

While each error that occurred might have been excused, taken as a whole they indicate poor methods of working and a lack of managerial auditing, by no means unusual at the time the accident occurred. In particular:

- The unsatisfactory state of the nitrogen supply was known and improvements were being studied but perhaps not with sufficient urgency. More thought might have been given to the prevention of contamination.
- There should have been better control of trip set-points.

References

This chapter is based on an article which was published in *Hydrocarbon Processing,* Vol. 59, No. 11, Nov. 1979, p. 373, and thanks are due to Gulf Publishing Co. for permission to quote from it.

1. KLETZ, T. A. *What went wrong? Case Histories of Process Plant Disasters,* Gulf Publishing Co., 1985, Section 18.2.
2. As ref. 1, Section 12.3.
3. KLETZ, T. A. *Plant/Operations Progress,* forthcoming.

Chapter 3

Poor procedures and poor management

A crude oil distillation unit was being started up after a major turnaround. It had taken longer than expected so stocks of product were low and it was important to get the unit on-line as soon as possible. The manager therefore decided to be present throughout the night so that he could deal promptly with any problems that arose. Perhaps also his presence might discourage delay. He was a young graduate who had been in the job for only a year, and an additional reason for being present was to see for himself what happened during a major start-up.

The distillation column was warming up. It had been washed out with water before the shutdown and the water left in the column had distilled into the reflux drum and had half filled it. There was a layer of light oil containing some liquefied petroleum gas (LPG) on top of the water. Reflux had not been started.

The foreman asked an operator to start up the water pump. (Some water was always produced and was pumped to a scrubber where impurities were removed; after washing out the column the production of water was greater than usual.) It was then discovered that a slip-plate (spade) had been left in the suction line to the pump on the drum side of the isolation valve (Figure 3.1). All the branches on the drum had been slip-plated during the turnaround to isolate the drum for entry. The other slip-plates had been removed but this one had been overlooked.

The manager estimated that shutting down the reboiler furnace, allowing it to cool, fitting a flex to the spare branch on the reflux drum, draining the contents to a safe place, removing the slip-plate and warming up again would result in 24 hours delay. A foreman fitter, a man of great experience, offered to break the joint, remove the slip-plate and remake the joint while the water ran out of it. He could do it, he said, before all the water ran out and was followed by the oil; he had done such jobs before.

After some hesitation the manager agreed to let the fitter go ahead. He dressed up in waterproof clothing and, watched by the process team, unbolted the joint and removed the slip-plate while the water sprayed out. Unfortunately he tore one of the compressed asbestos fibre gaskets, half of it adhering to one of the joint faces. Before he could remove it and replace it, all the water ran out and was followed by the oil.

Some of the LPG flashed as the oil came out of the broken joint. This cooled the joint and some ice formed, making it impossible to remake the joint. The fitter abandoned the attempt to do so.

The reboiler furnace was only 30 metres away. As soon as the oil appeared, one of the process team pressed the button which should have shut down the burners. Nothing happened. The process team had to isolate the burners one by one while the oil and vapour were spreading across the level ground towards the furnace. Fortunately, they did so without the vapour igniting.

Afterwards it was discovered that the protective system on the furnace had given trouble a day or two before the turnaround started. The process foreman on duty therefore took a considered decision to bypass it until the turnaround, when it could be repaired. Unfortunately there was so much work to be done during the turnaround that this late addition to the job list was overlooked.

Although there was no injury or damage, both could easily have occurred, and there was a day's loss of production. The incident was therefore thoroughly investigated, as all dangerous occurrences and near-misses should be. In the following we assume that each member of the five-man investigating panel wrote a separate report, each emphasizing different causes and making different recommendations. Of course, it was not really as tidy as this. The first two sets of recommendations were made at the time. The others were made later when the incident was selected for discussion in the manner described in the Introduction.

The five sets of recommendations are not alternatives. All are necessary if a repeat of the accident is to be prevented.

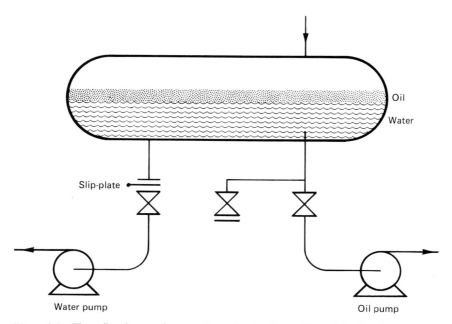

Figure 3.1   The reflux drum and connections, showing the position of the slip-plate

Report 1 – The slip-plate is the key

The author pointed out that the incident was entirely due to the failure to remove the slip-plate before the start-up. If the slip-plate had been removed the incident would not have occurred. He made the following recommendations which, like all those made in the other reports, apply generally and not just on the plant where the incident occurred.

- All slip-plates should be listed on a permit-to-work. It is not sufficient to say 'Slip-plate (or de-slip-plate) all branches on reflux drum'. Instead, their positions should be listed and identified by numbered tags.
- In addition, all slip-plates inserted during a shutdown should be entered on a master list and a final inspection made, using this list, before start-up.
- If the slip-plate had been inserted below the isolation valve it would have been possible to remove it with the plant on-line. Nevertheless, we should continue to insert slip-plates on the vessel side of isolation valves, as if they are fitted on the far side liquid might be trapped between the slip-plate and a closed valve and then slowly evaporate while people are working in the vessel. Such incidents have occurred.

Chapter 5 describes an accident which was the result of failure to *insert* a slip-plate.

Report 2 – Better control of protective systems

The author saw the failure to shut down the furnace as the key point. Leaks for one reason or another are inevitable from time to time and we must be able to rely on our protective equipment. He recommended that:

- Protective equipment should not be bypassed or isolated unless this has been authorized in writing by a responsible person.
- If it is bypassed or isolated this should be signalled to the operators in some way, e.g. by a light on the panel.
- All trips should be tested after a major turnaround and all trips that have been repaired or overhauled should be tested before they are put back into service. If testing is necessary outside normal hours and cannot be carried out by the shift team, then the people who carry out the testing should be called into work. (Chapter 2 described another incident which occurred because a trip had been made inoperable.)

Report 3 – Don't rush

The author saw the key event as a rushed decision by the manager. Few problems on a large plant are so urgent that we cannot delay action for 15 minutes while we talk them over. If those concerned had paused for a cup of tea they would have realized that removing the slip-plate was more hazardous than it seemed at first sight and that there were other ways of avoiding a shutdown. I cannot vouch for the authenticity of Figure 3.2 but I agree with the message.

Removing the slip-plate was more hazardous than seemed at first sight because the pressure at the slip-plate, due to the head of liquid, was nearly

Figure 3.2 The notice may not be authentic but the message is sound

10 lbf/in^2 (0.7 bar) higher than the pressure in the reflux drum (a gauge pressure of about 15 lbf/in^2 (1 bar)). This was not realized at the time. It might have been realized if those present had given themselves time to talk over the proposed course of action.

It is a useful rule of thumb to remember that if a column of liquid in a pipeline is x metres tall, it will spray out x metres horizontally if a flanged joint is slackened.

A shutdown could have been avoided with less risk by freezing the water above the slip-plate with dry ice and methanol or by injecting water into the reflux drum via the spare branch so as to maintain the level. Another possible way of avoiding the shutdown would be to remove the pump, pass a drill through the valve and drill through the slip-plate. This method could, of course, only be used if the valve was a straight-through type.

As a general rule, when we have to decide between two courses of action, both of which have disadvantages, there are often alternative actions available which we have not thought of.

Report 4 – Who was in charge?

The author saw the accident as due to the failure of the young manager to stand up to the foreman fitter. The manager's situation was difficult. The foreman was a strong personality, widely respected as an experienced craftsman, old enough to be the manager's father, and assured the manager that he had done similar jobs before. It was 3 a.m., not the best time of day for decisions. The manager could not be blamed. Nevertheless, sooner or later every manager has to learn to stand up to his foreman, not disregarding their advice, but weighing it in the balance. He should be reluctant to overrule them if they are advocating caution, but more willing to do so if, as in this case, they advocate taking a chance.

The foreman fitter felt partly responsible for the non-removal of the slip-plate. This made him more willing than he might otherwise have been to compensate for his mistake by taking a chance. A more experienced manager would have realized this.

Report 5 – The climate in the works

This author went deeper than the other two. He saw the incident as due to a failure to give sufficient emphasis to safety throughout the organization. What would the works manager have said the next morning if he found that the start-up had been delayed? Would he have commented first on the low stocks and lost production or would he have said that despite the low stocks he was pleased that no chances had been taken?

The young manager was not working in a vacuum. His judgement was influenced by his assessment of his bosses' reactions and by the attitude to safety in the company, as demonstrated by the actions taken or remarks made in other situations. Official statements of policy have little influence. We judge people by what they do, not what they say. If anyone is to be blamed, it is the works manager for setting a climate in which his staff felt that risk-taking was legitimate.

Did the manager feel that he had been given, by implication, contradictory instructions (like the operators at Chernobyl; see Chapter 12): in this case to get the plant back on-line as soon as possible and, at the same time, to follow normal safety procedures? Junior managers and foremen often find themselves in this position. Senior managers stress the importance of output or efficiency but do not mention safety, so their subordinates assume that safety takes second place. They are in a 'no-win' situation. If there is an accident they are blamed for not following the safety procedures. If the required output or efficiency is not achieved they are blamed for that. Managers, when talking about output and efficiency, should bring safety into the conversation. *What we don't say is as important as what we do say.*

How, if at all, did the young manager's training in the company and at university prepare him for the situation in which he found himself? Probably not at all. Today, in the UK, all undergraduate chemical engineers receive some training in loss prevention though it is unlikely to cover situations such as that described.

Other comments

Was it wise to attempt to shut down the furnace, either automatically or manually? It is probably safer to keep a furnace operating if there is a leak of flammable gas or vapour nearby. The flame speed is much less than that of the air entering a furnace (unless hydrogen is present) so flashback from the furnace is very unlikely. On the other hand, if the furnace is shut down, the gas or vapour may be ignited by hot brickworks.

The ground should have been sloped so that any liquid spillage flowed away from the furnace. In general, spillages should flow away from equipment, not towards it. (See Chapter 5.)

The incident illustrates the comment in the Introduction that accidents are rarely the fault of a single person but that responsibility is spread in varying degrees amongst many people: those who failed to remove the slip-plate, those who bypassed the furnace trip and then failed to make sure that it was repaired, the young manager, those responsible for his training and guidance, the foreman fitter, the works manager. Any of these, by doing their job better, had the power to prevent the incident. But

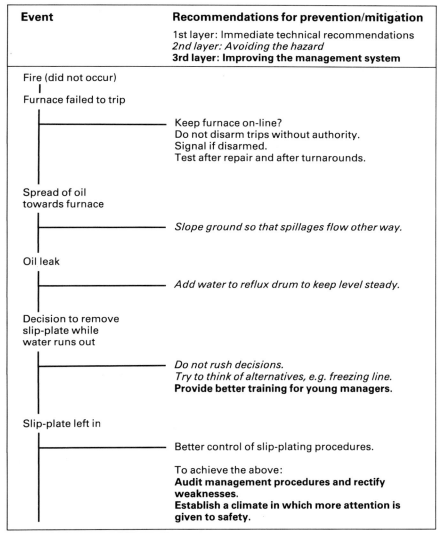

Event	Recommendations for prevention/mitigation
	1st layer: Immediate technical recommendations *2nd layer: Avoiding the hazard* **3rd layer: Improving the management system**
Fire (did not occur) Furnace failed to trip	Keep furnace on-line? Do not disarm trips without authority. Signal if disarmed. Test after repair and after turnarounds.
Spread of oil towards furnace	*Slope ground so that spillages flow other way.*
Oil leak	*Add water to reflux drum to keep level steady.*
Decision to remove slip-plate while water runs out	*Do not rush decisions.* *Try to think of alternatives, e.g. freezing line.* **Provide better training for young managers.**
Slip-plate left in	Better control of slip-plating procedures. To achieve the above: **Audit management procedures and rectify weaknesses.** **Establish a climate in which more attention is given to safety.**

Figure 3.3   Summary of Chapter 3

the systems for controlling slip-plating and the disarming of trips were unsatisfactory to say the least. At the operating level, those concerned were following custom and practice, and the greater responsibility is therefore that of the works manager and his senior colleagues, who either failed to recognize the deficiencies in their procedures or failed to do anything about them.

Reference

This chapter is based on a paper which was published in *Plant/Operations Progress,* Vol. 3, No. 1, Jan. 1984, p. 1, and thanks are due to the American Institute of Chemical Engineers for permission to quote from it.

Chapter 4

A gas leak and explosion – the hazards of insularity

Four men were killed, several injured and a compressor house was destroyed when a leak of ethylene ignited. Figure 4.1 shows some of the damage. Explosion experts estimated that between 5 and 50 kg of ethylene leaked out into the building during the eight minutes or so that elapsed between the start of the leak and the explosion. Detailed examination of the wreckage enabled the source of the leak to be identified. It was a badly made joint on a small-bore line.

The recommendations made to prevent similar explosions fell into four groups:

- Ways of preventing leaks or making them less likely.
- Ways of minimizing their effects.
- Ways of reducing the probability of ignition.
- Underlying all these, ways of removing the managerial ignorance and incorrect beliefs that led to the accident.

Most of the recommendations apply to other plants handling hazardous materials, particularly flammable liquids and gases, and many apply to all process plants.

Preventing leaks

As already stated, the leak of ethylene occurred from a badly made joint on a small-diameter line, about ½ inch internal diameter. The joint was a special type, suitable for use at high pressures, and assembling it required more skill than is required for an ordinary flanged joint. Once the joint had been made it was impossible to see whether or not it had been made correctly. It would be easy to say that more care should be taken in joint-making but this temptation was resisted and the underlying reasons why the joint had been badly made were investigated. There were two reasons.

(1) At one time the assembly of high-pressure joints was carried out only by a handful of craftsmen who were specially trained and judged to have the necessary level of skill and commitment. This was resented by the other craftsmen and a change was made, all craftsmen being

Figure 4.1   Some of the damage caused by the explosion

trained to the necessary standard, or so it was thought, and the work carried out by whoever was available. Manpower utilization was improved. Unfortunately some of the newly trained men did not have the necessary skill, or perhaps did not understand the importance of using their skill to the full, and the standard of joint-making deteriorated. Following the explosion a return was made to the original system.

In addition, several other actions were taken:

(a) A standard for the quality of joints was specified.

(b) An attempt was made to explain to the craftsmen why such a high quality of joint-making was necessary and the results that would follow if joint-making was not up to standard. Obviously this task was made easier by the results of the explosion, which were known to all concerned, and it would have been more difficult to do the same on a plant which had not suffered a serious leak. Nevertheless, it is something that should be attempted on many plants.

(c) A system of inspection was set up. The joints had to be inspected after the surfaces had been prepared and before the joint was made. Initially all joints were inspected but as the inspectors gained confidence in the ability of individual craftsmen they inspected only a proportion, selected at random. Obviously, in the atmosphere that existed after the explosion, the craftsmen accepted inspection more readily than they would have done at other times.

(d) Better tools, to make joint-making easier, were developed and made available.

These actions reduced the leak frequency by a factor of about 20.

(2) Although leaks were quite common, about ten per month, mostly very small, nobody worried about them. The attitude was 'They can't ignite because we have eliminated all sources of ignition'. Unfortunately, this view was incorrect. It is almost impossible to eliminate all sources of ignition and, as we shall see later, not everything that could have been done to eliminate them had been done. A study made after the explosion showed that several ignitions had occurred in the past, on this and similar plants, but that nevertheless the probability of ignition was very low, only about one leak in ten thousand igniting. This low probability may have been due to the fact that the high-pressure gas dispersed by jet mixing. On most plants the probability of ignition of small leaks is probably between one in ten and one in a hundred. (For large leaks, greater than several tonnes, the probability of ignition is much higher, greater than one in ten and perhaps as high as one in two.)

Another reason for the large number of leaks was the large number of joints and valves. The plant consisted of several parallel streams, each containing three main items of equipment. Their reliability was not high and so, in order to maintain production, a vast number of crossovers and isolation valves was installed so that any item could be used in any stream. The money spent in providing all this flexibility might have been better spent in investigating the reasons why on-line time was so poor. Later plants, built after the explosion, had fewer streams and fewer crossovers.

Minimizing the effects of leaks

Although much can be done, and was done, after the explosion, to make leaks less probable, they are still liable to occur from time to time, and several actions were taken, to various extents, to minimize their effects. Taken together they constitute a defence in depth. If one line of defence fails, the next one comes into operation. They may be summarized by the words: detect, warn, isolate, disperse, vent.

Detect   Flammable gas detectors provide a cheap and effective means of giving early warning that a leak has occurred. Action can then be taken to isolate the leak, evacuate personnel, call the fire service in case the leak ignites etc. They should be installed on all plants where experience shows the leaks of hazardous materials are likely to occur, such as all plants handling liquefied flammable gases. It is particularly important to install them in buildings, as very small leaks indoors can cause serious damage. In the open air several tonnes are usually needed for an explosion but in a building a few tens of kilograms are sufficient. It is also important to install them in places where men are not normally present, as leaks may otherwise continue for a long time before they are detected.

In the present case we know that the leak was detected within four minutes of its start, as one man visited the area of the leak before it started and another man saw the leak four minutes later. The procedure for shutting down the plant had been started but had not progressed very far when ignition occurred after a few minutes.

The compressor house where the leak occurred was on two floors. The compressors were located on the upper floor and operators were normally present in this part of the building. The ground floor, where the leak occurred, housed various ancillary equipment and was normally visited about twice per hour. It was therefore a fortune chance that the leak was detected so soon, and gas detectors were installed – on both floors – when the plant was rebuilt.

Gas detectors of an appropriate type should also be installed on plants handling toxic gases or vapours, particularly if experience shows that leaks are liable to occur.

Warn   Three of the men killed by the explosion, and most of the injured men, were maintenance workers who were repairing a compressor. Although only a few minutes elapsed between the discovery of the leak and the ignition, there was ample time for them to leave the building. No-one told them to do so as no-one considered it possible that the leak would ignite. It was not normal practice to ask people to leave the building when a leak occurred. In fact, very small leaks were often ignored until it was convenient to repair them.

If a leak of flammable (or toxic) gas is detected, by people or by gas detectors, all people who are not required to deal with the leak should leave at once. There should be a recognized alarm signal, actuated automatically or by a responsible person, and a known assembly point, not too near the plant.

Isolate   The fourth man killed was an operator who was attempting to shut down the plant. He at least had a reason for staying in the building when the leak was discovered. However, people should never be asked to enter or remain near a cloud of flammable gas in order to isolate a leak. Remotely operated emergency isolation valves should be provided. We cannot install them on the lines leading to and from all equipment which might leak but we can install them (a) when experience shows that leaks are liable to occur or (b) when, if leaks should occur, a large inventory will leak out. Examples of (a) are certain high-pressure compressors, as in the present case, and very hot or cold pumps. Examples of (b) are the bottom pumps on distillation columns[1, 2, 3].

In some cases, as well as remotely operated valves for isolating leaks, we also need remotely operated blow-down valves for getting rid of the inventory of hazardous material present in the section of plant that has been isolated. It may also be necessary to provide remote means of isolating power supplies such as pump and compressor drives. If a pump or compressor is fitted with remotely operated isolation valves, operation of these valves should automatically isolate the power supply.

If a remotely operated valve is fitted in the suction line of a pump or compressor, a non-return (check) valve may be installed in the delivery line instead of another remotely operated valve. This is acceptable provided that the valve is inspected regularly. Non-return valves have a bad name with many engineers but this may be because they are never looked at. No piece of equipment, especially one containing moving parts, can be expected to function for the lifetime of a plant without examination and repair if necessary. (On nuclear plants repair is often impossible and equipment has to be designed for lifetime operation, but normal processing plants do not use such special equipment.)

However many emergency isolation valves we install, a leak may occur on a piece of equipment which is not provided with them. Is it acceptable for someone to enter a cloud of flammable gas to isolate such a leak? I would not like to say 'never'. I know of cases where a quick dash into a cloud has stopped a leak which might otherwise have continued for a long time. Anyone doing this should be protected by water spray and suitable clothing. However, we should try to avoid putting people into situations where they have this sort of decision to make, by generous provision of emergency valves. There were none on the plant we are discussing but they were installed when it was rebuilt. (Chapter 15, item 6, describes another occasion when a man entered a vapour cloud to isolate a leak.)

Disperse   'The best building had no roof and no walls!' Whenever possible, equipment handling flammable liquids and gases should be located in the open air so that small leaks are dispersed by natural ventilation. Not only is dispersion better but much more gas or vapour, as already stated, has to be present before an explosion occurs. A roof or canopy over equipment is acceptable but walls should be avoided.

On the plant were the explosion occurred it was believed that it would be difficult to maintain the compressors adequately if they were in the open, and a closed building was installed on the rebuilt plant. Other plants, however, have found that they can manage to operate and maintain compressors in the open air with no more than a canopy over them. Temporary walls and heaters, for use during maintenance, have been described[4].

The company owning the plant where the explosion occurred operated several other plants on the same site. Many of these had open compressor houses (though the compressors needed less attention than the ones we are discussing). When the explosion occurred, a new compressor house had just been completed on one of these plants. It was near a workshop and so, in order to reduce the noise level in the workshop, they departed from their usual practice and built a closed compressor house. When they saw the report on the explosion they pulled down the walls before the compressor house was commissioned. The noise problem was dealt with in other ways.

If, despite my advice, you build a closed compressor house, then forced ventilation is better than nothing but not nearly as good as the ventilation you get for free, even on a still day, in a building without walls.

Another method of dispersing leaks is by the use of steam or water curtains. Steam curtains have to be fixed but water curtains can be fixed or temporary. Temporary water curtains using fire service monitors have often been used to confine and disperse leaks and thus prevent them igniting.

Vent   If equipment handling flammable gases or liquids has to be installed in a building then it is possible to minimize damage, if an explosion occurs, by building walls of such light construction that they blow off as soon as an explosion occurs, before the pressure has built up to a level at which it causes serious damage. Such walls do not, of course, protect people working in the building, as these will be killed or seriously injured by the fire, and are therefore very much a second best choice. The walls have to be made of light plastic sheets fixed by special fastenings, which easily break. In one case, when the designers specified such walls, the construction engineer was surprised that such weak fastenings had been requested. He did not know the reason for them. He therefore substituted stronger fastenings. When an explosion occurred the walls blew off at a higher pressure than the designers intended and damage was greater than it should have been.

Other lines of defence are fire-protection measures, such as insulation and water spray, and fire-fighting. They will not have any effect if an explosion occurs but may reduce the damage caused by a fire.

Reducing the probability of ignition

The source of ignition was never identified with certainty but two possible causes were found. Since either of them might have been the cause, action had to be taken to try to eliminate both of them in the future.

Faulty electrical equipment   The plant where the explosion occurred was classified as Zone 1 (Division 1) and the equipment was of flameproof design. Inspection of the equipment after the explosion, including sections of the plant not damaged by the explosion, showed that the standard of maintenance of the electrical equipment was poor. Gaps were too large, screws were missing, glasses were broken, unused entry holes were not plugged. A first glance, looking at equipment at eye level, suggested that nothing much was wrong. When equipment that could only be examined from a ladder was looked at, however, much of it was found to be neglected.

The action taken was similar to that taken to improve the standard of joint maintenance, namely:

(1) It was found that many electricians did not understand the construction of flameproof equipment and did not realize the importance of correct assembly. Training courses were set up, on other plants as well as on the plant where the explosion occurred.
(2) A system of regular inspection was set up. The inspectors examined a proportion of the equipment in use, selected at random, and some of the items inspected were dismantled. Similar inspections on other plants showed a sorry state of affairs, up to half the equipment being

found faulty, though not all the faults would have made the equipment a source of ignition. Reader, before you criticize, have you examined your equipment?

(3) It was found that the special tools and bolts required for assembling flameproof equipment were not in stock. If an electrician lost a tool or dropped a bolt, what was he expected to do?

(4) On some plants, though not to a great extent on the plant where the explosion occurred, it was found that flameproof equipment was being used when Zone 2 equipment would be quite adequate. It requires much less maintenance than flameproof equipment and the cost of replacing flameproof equipment by Zone 2 equipment was soon recovered.

Static electricity   Various bits of loose metal were found lying about the plant, mainly bits of pipe, valves and scaffold poles left over from maintenance work. If these were electrically isolated, as many of them were, and were exposed to a leak of steam or ethylene, then a charge of static electricity might accumulate on them. When the charge reached a high enough level it could flash to earth, producing a spark of sufficient power to ignite an ethylene leak. Good housekeeping is essential, all bits of loose metal being removed. If they cannot be removed, e.g. because they are required for a forthcoming turnaround, they should be grounded.

Other sources   There was no evidence of smoking or other illegal sources of ignition. All employees recognized the need to eliminate smoking. There was no maintenance work going on in the ground floor area at the time of the explosion so a spark from a tool dropped by a maintenance worker on the concrete floor can be ruled out. The steam pipes in the basement were not nearly hot enough to ignite ethylene.

Incorrect knowledge

The explosion would not have occurred if the people who designed and operated the plant had realized that:

• Sources of ignition can never be completely eliminated even though we do what we can to remove all known sources. The fire triangle is misleading in an industrial context and instead we should say:

air + fuel → bang

• We should therefore do everything possible to prevent and disperse leaks.

These statements were widely accepted on other plants on the same site. Explosions and fires had occurred on these plants in the past and lessons had been learnt from them. But these lessons had not been passed on to the plant where the explosion occurred or, more likely, they were passed on but nobody listened. The plant staff believed that their problems were different. If you handle ethylene as a gas it is perhaps not obvious that you can learn anything from an explosion on a plant handling propylene as a liquid. Such an explosion had occurred in the same company, a few miles away, about 10 years earlier. The recommendations made, and acted upon, were very similar to those outlined above. But no-one on the plant where the later explosion occurred took any notice. The plant was a retreat – a group of people isolating themselves by choice from the outside world – but fortunately the explosion blew down the cloister walls. Not only did the staff adopt many of the beliefs and practices current elsewhere, even, in time, building open compressor houses, but they developed a new attitude of mind, a much greater willingness to learn from the outside world.

The explosion illustrates the words of Artemus Ward, 'It ain't so much the things we don't know that get us in trouble. It's the things we know that ain't so'.

Blinkered attitudes, however, continued elsewhere in the company. Before the explosion some low-pressure ethylene pipework had been constructed of cast iron. This suffered much greater damage than steel pipework and it was agreed that cast iron would not be used in future. Only steel was used in the rebuilt plant. A couple of years later, construction of a plant which used propylene was sanctioned. The design team insisted that rules agreed for ethylene did not necessarily apply to them, and the arguments against cast iron had to be gone through all over again.

Similarly, although the recommendations made after the explosion, particularly the need for open compressor houses, were circulated throughout the company, little attention was paid to them on plants that handled hydrogen. Hydrogen, it was agreed, is much ligher than any other gas and disperses easily. A few years later the roof and walls were blown off a hydrogen compressor house.

Individual parts of the company were allowed considerable autonomy in technical matters. Many years before it had been formed by an amalgamation of independent companies who still cherished their freedom of action and it was felt that attempts to impose uniform standards and practices would lead only to resentment. The explosion did not produce a re-examination of this philosophy though perhaps it should have done. It probably never occurred to the senior managers of the company that their organizational structure had any bearing on the explosion – it was due to a badly made joint. The management philosophy was not changed until many years later when recession caused different parts of the company to be merged.

However, a word of caution. Although the senior managers of the company might have prevented the accident by changing their organization and philosophy, do not use this as an excuse for doing less than possible at other levels. The chain of causation could have been broken at any level from senior manager to craftsman. The accident might not have occurred if the organization had been different. Equally it would not have occurred if the joint had been properly made.

Other explosions in closed buildings are described in references 5, 6 and 7.

Other points

Another point illustrated by the explosion is that although ethylene at high pressure and atmospheric temperature is a gas, being above its critical temperature, it behaves in some ways like a flashing liquid, i.e. a liquid under pressure above its normal boiling point. Thus its density is close to that of a liquid and a small leak produces a large gas cloud. The precautions which should be taken when handling it are similar to those that should be taken when handling liquefied flammable gases such as propylene. There was more to be learned from the propylene explosion 10 years earlier than the staff realized.

Postscript

After I had completed this chapter I came across a memorandum on 'Safety Precautions on Plant Handling Ethylene', issued thirty years before the explosion, by the predecessor company, when they started to use ethylene. It said: 'In all the considerations that have been given in the past to this question, it has always been felt that excellent ventilation is a *sine qua non*. Not only does it reduce the danger of explosive concentrations of gas occurring, but it also protects the operators of the plant from the objectionable smell and soporific effects of ethylene'.

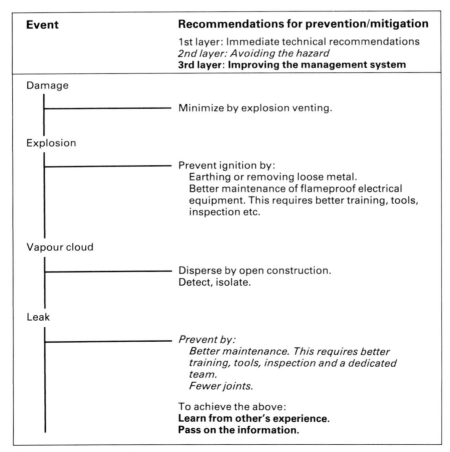

Event	Recommendations for prevention/mitigation
	1st layer: Immediate technical recommendations *2nd layer: Avoiding the hazard* **3rd layer: Improving the management system**
Damage	
	Minimize by explosion venting.
Explosion	
	Prevent ignition by: Earthing or removing loose metal. Better maintenance of flameproof electrical equipment. This requires better training, tools, inspection etc.
Vapour cloud	
	Disperse by open construction. Detect, isolate.
Leak	
	Prevent by: *Better maintenance. This requires better training, tools, inspection and a dedicated team.* *Fewer joints.* To achieve the above: **Learn from other's experience.** **Pass on the information.**

Figure 4.2   Summary of Chapter 4

During the thirty years this sound advice was forgotten and ignored. Chapters 5 and 13 describe other accidents which occurred because knowledge was lost.

References

1. KLETZ, T. A. *Chemical Engineering Progress,* Vol. 71, No. 9, Sept. 1975, p. 63.
2. KLETZ, T. A. *What went wrong? Case Histories of Process Plant Disasters,* Gulf Publishing Co., 1985, Section 7.2.
3. LEES, F. P. *Loss Prevention in the Process Industries,* Butterworths, 1980, Vol. 1, Section 12.6.2.
4. MORRIS, D. H. A. *Loss Prevention in the Process Industries,* edited by C. H. Buschmann, Elsevier, Amsterdam, 1974, p. 369.
5. HOWARD, W. B. *Loss Prevention,* Vol. 6, 1972, p. 68.
6. *The Fire and Explosions at Permaflex Ltd, Trubshaw Cross, Longport, Stoke on Trent, 11 February 1980,* Health and Safety Executive, London, 1981.
7. *Vigilance,* Vol. 4, No. 7, Summer 1987, p. 70.

Chapter 5

A liquid leak and fire – the hazards of amateurism

Two men were killed and several seriously injured when a leak of 4 tonnes of hot flammable hydrocarbon ignited. There was no explosion, only a fire which was soon extinguished. Material damage was relatively light – about £1M at 1986 values – but the consequential loss (business interruption) was 30 times this figure.

The hydrocarbon processing plant consisted of two parallel units which shared some equipment in common (Figure 5.1). One of the units was shut down for maintenance while the other continued on-line. The unit under repair should have been isolated by slip-plates (spades) but unfortunately one of the interconnecting lines was overlooked. As a result the leak occurred.

We shall therefore look at:

- The action needed to prevent similar leaks in the future.
- Ways of reducing the probability of ignition.
- Ways of minimizing the effects of the fire.
- The influence of the management experience and philosophy.

First, however, let us look at Figure 5.1 in a little more detail. The reaction sections of the two units were quite separate. Reaction product entered a common product receiver and was then further processed in a common unit. The product receiver could be bypassed by the lines shown dotted in Figure 5.1. These lines had been installed to make it easier to empty the reactors when they were being shut down and were used only occasionally. They were therefore overlooked when slip-plates were inserted to isolate the operating unit from the one that was to be maintained.

Valve B, in the bypass line from the shutdown unit (No. 2), had been removed for overhaul. A few hours later valve A was operated. Hot hydrocarbon (boiling point about 85°C, temperature about 150°C and gauge pressure about 7 bar) travelled in the wrong direction along No. 2 unit bypass line and came out of the open end. Someone sounded the evacuation alarm and the numerous maintenance workers who were on the plant at the time, working on No. 2 unit, started to leave. Unfortunately, for some of them the direct route to the main gate and assembly point lay close to the leak. As they were passing it, 90 seconds after the leak started,

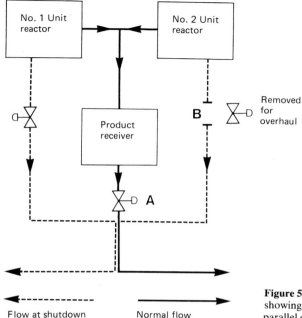

No. 1 Unit reactor

No. 2 Unit reactor

Product receiver

B

Removed for overhaul

A

Figure 5.1 Simplified flow diagram showing a combination of series and parallel equipment

Flow at shutdown   Normal flow

the hydrocarbon ignited. However, most of the 80 maintenance workers who were on the plant at the time escaped without injury.

Valve A had been operated remotely from the control room. There were no windows and the operator could not see the plant. He heard the evacuation alarm sound. He did not know the reason but as he had just operated valve A, he thought this might be connected in some way with the alarm and he shut it again. His prompt action prevented a much more serious fire.

Because the hydrocarbon was under pressure above its normal boiling point a very large proportion of the leaking material turned to vapour and spray. The fire was therefore much more extensive than that which would have followed the spillage of 4 tonnes of cold hydrocarbon. The initial flash fire covered a substantial area of the plant and was followed by a smaller fire close to the point of leakage.

Prevention of the leak

The policy on the plant where the fire occurred was that equipment under repair should be isolated by slip-plates (or disconnection of a section of pipework) unless the repair job was a quick one. If a whole unit or section of a unit was shut down for maintenance then that unit or section could be isolated as a whole and it was not necessary to isolate each individual item of equipment. This was a sound policy and the intention was to follow it, but unfortunately a little-used interconnecting line between the two units was overlooked. Although a permit-to-work was issued authorizing the maintenance organization to separate the two units, it did not list each joint to be slip-plated or disconnected but merely said, 'Insert slip-plates or disconnect joints to separate No. 1 unit from No. 2 unit'. The plant was not, of course, as simple as implied by Figure 5.1; it was a 'spaghetti bowl'.

Chapter 3 described an accident which occurred because one slip-plate was left in position when others were removed.

After the fire the following recommendations were made to prevent a similar incident in the future.

(1) A schedule of isolations should be prepared well in advance of a shutdown, using the line diagrams, and checked on the ground by tracing all interconnecting lines. Particular care should be taken when tracing lines that are insulated together. Line diagrams should be kept up-to-date.

(2) The joints to be slip-plated (or disconnected) should be marked with a numbered tag and listed individually on a permit-to-work. It is not sufficient to write on the permit, 'Fit slip-plates to separate No. 1 unit from No. 2 unit'.

(3) When two (or more) units are closely interconnected, extended maintenance should not be carried out on one of them while the other is operating. However, it may not always be possible to follow this recommendation on existing plants.

(4) When designing new plants, all equipment or sections of plant that will have to be maintained while the rest of the plant is on-line should be identified and a check made that they can be simply and safely isolated. This check is included in the hazard and operability studies (hazops)[1,2] which many companies now carry out on all new designs.

These recommendations apply to all plants handling hazardous materials and not just to the plant where the fire occurred.

The source of ignition

The source of ignition was an unusual one and attracted widespread interest at the time. The company concerned issued a press release and accounts appeared in several journals[3]. As a result little interest was shown, outside the company, in the reasons why the leak had occurred.

A diesel-engined vehicle was being used by the maintenance team who were working on the unit that was shut down. When the leak occurred, hydrocarbon vapour was sucked into the engine through the air inlet and the engine started to race. The driver tried to stop the engine, in the normal way, by isolating the fuel supply, but as the engine was getting its fuel through the air inlet it continued to race. Valve bounce occurred and the vapour cloud was ignited by flashback from the cylinders.

The company concerned did not allow petrol engines into areas where flammable gases or liquids were handled but diesel engines were considered safe as they produce no sparks. The fire showed that they are as dangerous as petrol engines and should be treated with the same respect. In fact, several other leaks had been ignited by diesel engines in the same way, and in other ways, but they had received little publicity.

Since the time of the fire several proprietory devices have been devised which make it possible to shut down a diesel engine which is sucking in fuel through the air inlet. A valve in the air inlet closes and in addition a fire-extinguishing agent such as carbon dioxide or Halon is injected into the air inlet. The devices can be operated by hand or they can be operated automatically by a combustible gas detector.

Diesel engines can ignite leaks in several other ways and if we wish to prevent them doing so the exhaust temperature should be kept below the auto-ignition temperature of the gas or vapour, a spark arrester and flame arrester should be fitted to the exhaust and the decompression control should be disconnected. The electrical equipment should be suitable for Zone (Division) 1 or 2 areas.

The following recommendations, of general applicability, were made after the fire:

(1) Diesel engines which operate for more than 1000 hours per year in Zone (Division) 1 or 2 areas should be fully protected with a device that enables them to be stopped if fuel is being sucked in through the air inlet, spark and flame arresters on the exhaust and suitable electrical equipment. Compressed air or spring starters should be used instead of electric starters. The exhaust temperature should be kept below the auto-ignition temperature of the materials which might leak and the decompression control should be disconnected. Fixed diesel engines should whenever possible be located in safe areas or, if this is not possible, their air supplies should be drawn from safe areas and their exhausts should discharge to safe areas, through long lines if necessary.

(2) Vehicles which are used occasionally, e.g. during maintenance operations, can be protected to a lower standard if they are never left unattended. They should, however, be fitted with a device which enables them to be shut down if fuel is being sucked in through the air inlet, the exhaust temperature should be below the auto-ignition temperature of the materials which might leak and the decompression control should be disconnected. If there are no combustible gas detector alarms in the area then portable ones should be installed.

(3) Vehicles which are just passing through, e.g. delivering or collecting materials, need not be protected at all but should not enter without permission from the process foreman. This permission should not be given unless conditions are steady and there are no leaks.

The effect of these recommendations was to discourage the use of cranes, welding sets etc. in Zone (Division) 1 and 2 areas while plants are on-line, while falling short of stopping it completely. When cranes were used, the use of hydraulic cranes in place of diesel electric cranes was encouraged.

Minimizing the effects of the fire

The plant was a 'No Smoking' area and was therefore surrounded by a fence to control access. Only one gate was in regular use. As already stated, the direct route to this gate and the prearranged assembly point, for some of the 80 maintenance workers, lay close to the leak. When the evacuation alarm sounded these men started to leave and were passing close to the leak when it ignited. Other men left in other directions by climbing the fence.

After the fire the rules were changed. Additional gates, which could be opened only from the inside, were installed and employees were told to leave by the nearest gate and make their way round the plant, if necessary, to the assembly point.

The fire drew attention to the fact that 80 men, in addition to the normal operating team, were working on a plant that was in partial operation. This is bad practice, as already stated, and plants should not be designed with closely interconnected parallel units if they are to be overhauled at different times.

The congested design of the plant made rapid evacuation difficult and increased the damage caused by the fire. In addition the ground was sloped so that spillage flowed towards the centre of the plant instead of away from it (as in the accident described in Chapter 3). Much of the drainage flowed along open channels instead of underground drains. Both these features encouraged the spread of fire, and considerable sums were spent afterwards in trying to alleviate them. It would have cost little more at the design stage if the ground had been sloped so that spillage ran away from the main items of equipment and drains were underground.

As already stated, the control room operators could not see the plant but they heard the evacuation alarm; one of them assumed that his last action might have been responsible, and reversed it. This prompt action prevented a more serious fire. The probability that an operator will take the correct action in such a situation may be increased if he can see the plant through the control room windows. Windows are a weak spot if a control room is being strengthened to resist blast but nevertheless many people feel that they are worth having in order to give the operators a view of the plant. In theory, all the information they need to control the plant can be obtained from instruments or closed circuit television. In practice, many operators feel that a view of the plant is helpful when things are going wrong. Windows need not be very big and should be made of shatter-resistant glass or glass protected with plastic film.

The management experience and philosophy

If the explosion described in Chapter 4 was due to a blinkered attitude – deliberate isolation from the rest of the organization – the fire just described was due to amateurism. The company involved had been engaged in batch chemicals production for many decades and were acknowledged experts. Sales of one product, which had started as a batch product, had grown so much that construction of a large continuous plant was justified; in fact, several such plants had been built. The company had little experience of continuous production so they engaged a design contractor and left it to him. Unfortunately the contractor they chose was able to offer a good process but a poor engineering design. The plant was extremely congested, thus aggravating the effects of the fire, and the combination of single stream and parallel operation, with the two streams closely interconnected, both physically and in the line-diagram sense, made maintenance difficult and increased the likelihood of error. The drainage arrangements were almost scandalous.

The company was part of a larger group, many of the other companies in the group being involved in the design and operation of continuous plants. They had learnt by bitter experience that they should not engage a contractor, however competent, and leave the design to him. They monitored the design closely as it developed. They had found that employment of a contractor did not reduce the amount of graduate engineer effort required, though it did, of course, vastly reduce the draughtsman effort required. At no time did the batch company seek help or advice from the other companies in the group. It probably did not even occur to them that they might usefully seek advice. Like the unfortunate

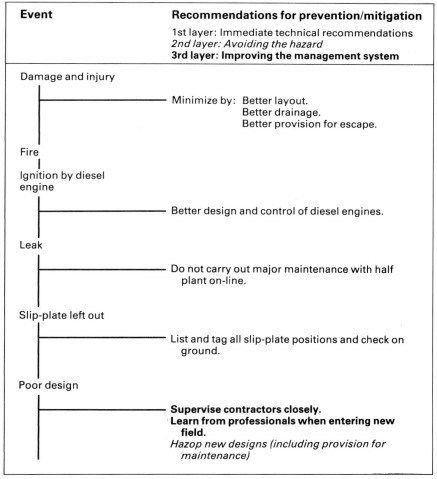

Figure 5.2 Summary of Chapter 5

engineer at Flixborough who designed the pipe which failed (Chapter 7), they did not know what they did not know. The managers of the holding company allowed its component parts considerable autonomy in technical matters and never considered that they should comment on choice of contractor or say who should be consulted. However, about two years after the fire there was a major reorganization, and responsibility for the plant involved in the fire, and other similar plants, was transferred to another part of the group.

As with the accident described in Chapter 4, it probably never occurred to the senior managers of the company that their actions, or inactions, had contributed towards it, and that if they had taken advice from those more experienced in handling large continuous plants it might never have occurred. But again the chain could have been broken at any point. The accident would not have occurred if the senior managers had been less insular; equally, it would not have occurred if the foreman responsible for seeing that the slip-plates were fitted had inspected the plant more thoroughly.

References

1. KLETZ, T. A. *Hazop and Hazan – Notes on the Identification and Assessment of Hazards,* Institution of Chemical Engineers, Rugby, UK, 2nd edition, 1986.
2. LEES, F. P. *Loss Prevention in the Process Industries,* Butterworths, 1980. Vol. 1, Section 8.9.
3. For example, *Chemical Age,* 12 Dec. 1969, p. 40 and 9 Jan. 1970, p. 11.
4. OIL COMPANIES MATERIALS ASSOCIATION, *Recommendations for the Protection of Diesel Engines Operating in Hazardous Areas,* Wiley, 1977. A British Standard is in preparation.

Chapter 6

A tank explosion – the hazards of optional extras

An explosion followed by a fire occurred in a 1000 m³ fixed roof storage tank which was one-third full of a volatile hydrocarbon (Figure 6.1). The roof was blown off the tank but remained attached at one point. Application of the theoretical quantity of foam from hoses failed to extinguish the fire, probably because the foam dried out before it reached the surface of the burning liquid. However, a monitor delivering 5000 gallons/min (23 m³/min) of foam extinguished the fire in 10 minutes. Nobody was injured. The hydrocarbon (flashpoint 45°C) was unconverted raw material recovered from a product by distillation and was contaminated with 0.5% of a more volatile liquid which lowered the flashpoint to near-ambient. The contaminant had a high conductivity. The

Figure 6.1 The inside of the tank after the fire

tank was being used on balance, liquid being added to the tank and
withdrawn for recycling at about the same rate $(2\,m^3/hr)$.

For a fire we need fuel, air (or oxygen) and a source of ignition, and we
shall consider these separately before looking at the weaknesses in the
management systems. It is also useful to ask, when considering a fire or
explosion, or indeed any accident, why it happened when it did and not at
some other time.

The fuel

Despite the contamination of the liquid in the tank with 0.5% of a more
volatile material, its flashpoint was still above the ambient temperature.
However, as a result of an upset on the distillation column the amount of
contaminant in the incoming stream rose to 9% and the flashpoint fell. If
this incoming liquid had been added to the base of the tank, the usual
practice when adding liquid to a tank, the change in the composition of the
liquid would have had little effect on the composition of the vapour. But
the tank was fitted with a swing arm so that liquid could be withdrawn from
any level. This swing arm was stuck in the fully raised position and could
not therefore be used for withdrawing liquid from the one-third full tank. It
was therefore being used for delivery, and the normal delivery line into the
base of the tank was being used for suction (Figure 6.2). As a result the

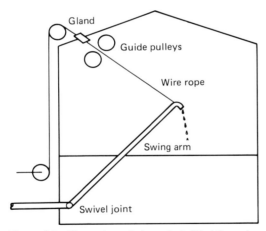

Figure 6.2 The tank was being splash-filled through a swing arm which was stuck in the
raised position. (Drawing not to scale)

composition of the vapour space reflected the composition of the
incoming, splashing liquid rather than the composition of the bulk of the
liquid in the tank, and the vapour composition changed as soon as the
composition of the incoming liquid changed. This probably explains why
the explosion occurred soon after the upset on the distillation column.

The unit manager knew that it was bad practice to splash-fill a tank but
he felt he had no other option, and he did not consider the risk serious, as
he thought the tank was blanketed with nitrogen.

The air

It was company policy to blanket with nitrogen all fixed roof storage tanks containing hydrocarbons with flashpoints below or close to ambient temperature, and the tank which exploded should have been blanketed. The system used was old-fashioned but quite effective it used correctly. All the fixed roof storage tanks were connected to a gasholder. When liquid was added to a tank, nitrogen was pushed into the gasholder; when liquid was withdrawn from a tank, nitrogen was withdrawn from the gasholder. This system conserved nitrogen but today the value of the saving would not justify the cost of installing a gasholder. Customs and Excise regulations required the tanks to be manually dipped every month and for this to be done they had to be disconnected from the gasholder and vented to atmosphere for a short time, otherwise nitrogen would be wasted and would also be discharged from the dip hole close to the dipper.

After the explosion it was found that the tank was disconnected from the gasholder and open to the atmosphere. The unit manager had personally inspected the tank seven months before the explosion, when it had been first used on its current duty, and at the time the valves on the tank had been set correctly. It is believed that during the intervening months the operator who dipped the tank had failed to reconnect it to the nitrogen system and that thereafter other operators had left it as they found it. No operator would admit that he found or left the valves in the wrong position. The manager and foremen had not looked at them during the seven months period.

It was perhaps as well that the tank was disconnected from the nitrogen blanketing system, as the nitrogen in the gasholder was found to contain 15% oxygen. If the tank had been connected to the gasholder and other tanks then the explosion might have spread through the connecting lines to every tank on the site. For once, two wrongs did make a right!

How did the air get into the nitrogen system? It is believed that it leaked in through tank sample and dip holes which had been left open and through an open end which was discovered on one of the storage tanks. In addition it is possible that the gasholder may have been allowed to get down to its minimum position, in which case air would have been sucked into the tanks through their pressure/vacuum valves (conservation vents).

After the explosion a special tank inspector was appointed to check the position of the nitrogen valves on every storage tank every day. He carried a portable oxygen analyser and also checked the composition of the atmosphere in each tank. There were many tanks on the site and it was considered that the foremen would not have time to personally inspect them regularly. In addition, it would have been a good idea to have installed a permanent oxygen analyser alarm on the gasholder.

The source of ignition

Several sources of ignition were considered and eliminated. At first, static electricity was considered to be the obvious source but the filling rate was so low and the conductivity of the liquid so high that sufficient charge could not accumulate. A discharge from a mist of oil droplets was considered but an experiment (with water) showed that liquid falling the distance that it actually fell at the actual flow rate ($2\,m^3/hr$) did not produce any significant amount of mist. The walls of the tank were examined for traces of pyrophoric or catalytically active material and the liquid was examined for traces of peroxides but nothing was found. External sources of ignition could be ruled out, as there was no welding or other hot work in progress anywhere near the tank at the time, the weather was fine and the tank was located in a 'No Smoking' area.

A possible source of ignition that could not be ruled out is based on an observation by the Safety in Mines Research Establishment[1] that when a wire rope is subjected to friction small 'hairs' rubbed off the wire become incandescent. These hairs will not ignite methane but it is possible that they could ignite the volatile hydrocarbon or the contaminant, both of which have a much lower auto-ignition temperature than methane. The swing-arm, which, as already stated, was stuck in the upright position, was supported by a wire rope which passed over a pulley inside the tank. The pulley was seized and the wire rope had been tightened to try to reduce the vibration of the swing-arm. It is therefore possible that the heat or frictional 'hairs' produced by the vibration of the rope against the pulley provided the source of ignition.

The incident confirms the statement made in Chapter 4 that possible sources of ignition are so numerous that we can never be certain that we have eliminated them completely, even though we try to remove all known sources. Flammable mixtures should never be tolerated except in a few special cases when the risk of ignition is accepted. One such case is the vapour spaces of fixed roof storage tanks containing liquids of high conductivity such as alcohols and ketones, as there is little or no chance of ignition by static electricity, if the tanks are earthed[2].

The company concerned required fixed roof tanks containing volatile hydrocarbons to be blanketed with nitrogen, as many explosions in such tanks have been reported in the literature[3], but did not require blanketing on tanks containing high-conductivity liquids, as very few explosions have been reported in such tanks. Although the tank which exploded contained hydrocarbon, the contaminant gave it a high conductivity. Should the policy therefore be revised? The company decided to continue with its policy but to make it clear that splash filling and any form of mechanical movement or vibration could not be accepted.

An unlikely coincidence?

As in the accident described in Chapter 2, at first sight the explosion seems to have been the result of an unlikely coincidence: the loss of nitrogen blanketing, the rise in concentration of the volatile component and the source of ignition all occurred at the same time. Who could have foreseen that? In fact, the scene was set by three ongoing unrevealed faults, and when a fourth occurred, the explosion was inevitable. The three ongoing faults were:

• The nitrogen blanketing was out of use and probably had been for months.

- Splash filling had been in use for months.
- The source of ignition had probably been present for some time.

When the concentration of the volatile component in the inlet stream rose, as it was likely to do sooner or later, the splash filling ensured that the concentration in the vapour space also rose, and an explosion was inevitable.

Limitation of damage

Fortunately the tank was provided with a weak seam roof. The roof/wall weld failed first, as it was supposed to do, and there was no spillage. The fire burned harmlessly in a confined space and caused little damage. If the tank had failed elsewhere, burning liquid would have filled the bund which was shared with other tanks. The incident therefore shows the value of weak seam roofs.

A flare stack was located near the tank. Two guy ropes supporting this flare stack passed over the tank. The lower rope, which was close to the tank, was broken by the roof when it blew off. Fortunately the upper rope, though exposed to the heat, did not fail. If it had, the stack might have collapsed. Storage areas and operating plant should be better segregated.

Training and inspections

The explosion would not have occurred if the nitrogen blanketing system had been kept in working order. For any protective system to be kept in working order, those who operate it must be convinced of its value and regular checks must be made to make sure that it is in working order. These two actions go together. If managers do not make regular checks, then everyone assumes that it cannot be very important.

It is clear that operators, foremen and perhaps junior managers did not regard nitrogen blanketing as a major safety precaution. It was looked upon as an 'optional extra', to use a phrase from motor car catalogues: a luxury extra to adopt if you can but something that can be dropped if you are busy or nitrogen is scarce. In contrast, everybody regarded the prohibition of smoking as an important safety measure and there was no difficulty enforcing it. Incidents of smoking were almost unknown and the few that had occurred were confined to contractors, not regular employees.

The nitrogen blanketing system had been installed many years before. It had fallen into disuse but five years before the explosion a survey of tank explosions had been made and a decision taken to bring the nitrogen blanketing back into use. The necessary instructions were issued but no attempt was made to explain to employees, at any level, why the change had been made. We do not live in a society in which management by edict is possible. People want to know why they are asked to do things, particularly when they are asked to change their habits. The management had failed to provide this explanation. A written explanation would not have been sufficient. Any such change in policy should be discussed with those who will have to carry it out and their comments obtained.

But explanations are not enough. They must be followed by regular audits or inspections to make sure that the new practices are being followed. These also the management had failed to provide. After the explosion, as already stated, they appointed a full-time tank inspector, as there were so many tanks to be checked, and managers themselves made occasional checks.

This incident has been discussed on many occasions with groups of managers and design engineers, using the method described in the Introduction. Sometimes they comment on the rather old-fashioned nature of the nitrogen blanketing system and suggest a more modern once-through system. However, the old-fashioned system was quite safe if used properly. If people did not use the equipment provided, would they have used better equipment? Nevertheless, a once-through system is better as each tank is then isolated and if anything goes wrong with the blanketing on one tank it does not spread to the other tanks. Reference 4 describes an explosion in a pair of tanks which shared a common system.

At Bhopal (Chapter 10) much of the safety equipment was out-of-use.

As already described, the tank had to be disconnected from the nitrogen system every month so that it could be dipped. This was tedious and so there was a temptation for operators to leave the tank disconnected ready for the next month. If the safe method of operation is more troublesome than the unsafe method then many people will use the unsafe one. Systems should be designed so that safe operation is not difficult. If a dip-pipe had been installed in the tank it would have been possible to dip it without disconnecting it from the nitrogen system.

Avoiding the problem

Was it necessary to store the recovered hydrocarbon or to store so much? In the original design it was recycled as it was produced and there was no intermediate storage. The unit manager decided that buffer storage would make operation easier, and as a spare storage tank was available he made use of it. He probably gave no thought to the increase in hazard or the minimum quantity he could manage with. Storage of flammable liquids in tanks was an accepted hazard; there was only a small increase in the total quantity on site, and if the manager had not used the tank for this purpose then somebody else would have put something else in it. Nevertheless large stocks are made up of smaller stocks, and the need to keep storage to the minimum (for economic reasons as well as safety ones) should always be emphasized. 'What you don't have, can't explode.'

The accident at Bhopal in 1984, which killed 2000 people (Chapter 10), was also due to unnecessary storage of a hazardous intermediate.

Seeing the signal above the noise

This explosion occurred in the 1960s. Nevertheless all the lessons that can be learned from it, described above, are still relevant. Many accidents still occur because protective equipment is neglected or hazards are not explained to those who have to control them. These lessons were learnt in the company concerned. In addition the incident had a wider effect. The company had a good safety record and attached great importance to safety but it was considered a non-technical subject. The safety officers were mainly non-technical and concerned themselves with mechanical hazards, protective clothing and similar matters. The explosion was the first of a series of accidents which were to occur in the next few years and which gradually led the management to realize that there was a need for a technical input to safety, for a new sort of safety adviser who understood

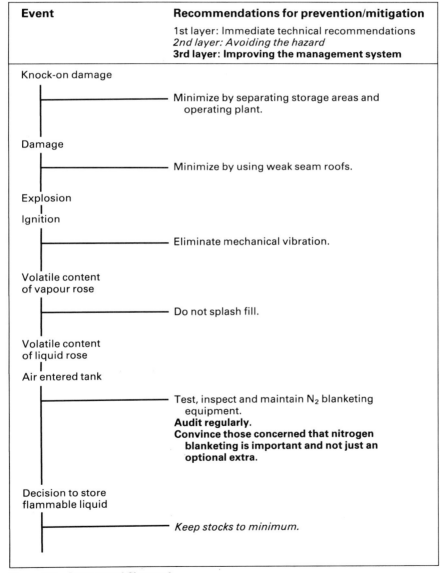

Event	Recommendations for prevention/mitigation
	1st layer: Immediate technical recommendations *2nd layer: Avoiding the hazard* **3rd layer: Improving the management system**
Knock-on damage	
	Minimize by separating storage areas and operating plant.
Damage	
	Minimize by using weak seam roofs.
Explosion Ignition	
	Eliminate mechanical vibration.
Volatile content of vapour rose	
	Do not splash fill.
Volatile content of liquid rose Air entered tank	
	Test, inspect and maintain N_2 blanketing equipment. **Audit regularly.** **Convince those concerned that nitrogen blanketing is important and not just an optional extra.**
Decision to store flammable liquid	
	Keep stocks to minimum.

Figure 6.3   Summary of Chapter 6

the technology and could advise on the action needed to prevent the sort of accidents which were now occurring, accidents similar to those described in Chapters 4, 5 and 7.

What had changed? This is a matter of conjecture but the following factors were probably important. First, during the 1960s a new generation of plants had been built which were larger than those built before and operated at higher temperatures and pressures; leaks, if they occurred, were therefore more serious. Second, plants were more dependent on instrumented protective systems but managers were slow to realize the amount of testing and maintenance that was required. Most of the operators had grown up on an earlier generation of plants and were slow to adapt to new ways. Managers did not realize the amount of retraining that was required.

Whatever the reasons for the changes, there is no doubt that it took several years and several serious fires and explosions before it was generally realized in the company that there had been a step change. Each fire and explosion was followed by a series of recommendations designed to remove the immediate technical causes, and to improve such matters as testing and inspection, but it was several years before it was realized that an entirely new approach, the loss prevention rather than the traditional safety approach, was needed. It is always difficult to see the signal above the noise when we do not know what signal to expect and we hope there is no signal at all.

References

1. POWELL, F. and GREGORY, R. *Technical Report No. 41,* Safety in Mines Research Establishment, Sheffield, UK, 1967.
2. KLETZ, T. A. *Loss Prevention in the Process Industries, Symposium Series No. 34,* Institution of Chemical Engineers, Rugby, UK, 1971, p. 75.
3. KLINKENBERG, A. and VAN DER MINNE, J. L. *Electrostatics in the Petroleum Industry,* Elsevier, Amsterdam, 1958.
4. KLETZ, T. A. *What went wrong? Case Histories of Process Plant Disasters,* Gulf Publishing Co., 1985, Section 5.4.2.

Chapter 7

Another tank explosion – the hazards of modifications and ignorance

One end of a horizontal cylindrical tank (volume $20\,m^3$) was blown off, killing two men who were working nearby (Figure 7.1). The tank was used to store a product of melting point 100°C which was handled as a liquid and kept hot by a steam coil. The gauge pressure of the steam was $100\,lbf/in^2$ (7 bar). The line into the tank was steam-traced and insulated but any faults in the steam tracing or insulation resulted in the formation of plugs of solid; the line had to be dismantled on several occasions to clear them. It was

Figure 7.1 The end blown off the tank by gas pressure

55

therefore emptied by blowing with compressed air after it had been used, and it was again blown with compressed air, to prove that it was still clear, before it was used again.

The accident happened during this blowing operation. An operator connected an air hose to the transfer line at the production unit, about 200 metres away, and then went to the tank to see if he could hear the air coming out of the vent – an open hole 3 inches (75 mm) in diameter. The air was not getting through. He spoke to two maintenance workers who were working nearby on a job unconnected with the tank. He then went to report to his foreman that no air was coming out of the tank vent and that the transfer line was presumably choked, not for the first time! While he was talking to the foreman they heard a bang, and going outside found that one end of the tank had blown off, killing the two maintenance workers.

It was later found that the vent was plugged with solidified product. The tank was designed to withstand a gauge pressure of 5 lbf/in^2 (0.3 bar) and would burst at a gauge pressure of about 20 lbf/in^2 (1.3 bar), so in retrospect it is not surprising that the plug of solid was stronger than the tank.

In discussing this accident it will be convenient to deal with each aspect – the control of plant modifications, the choking of the vent, knowledge of material strength and learning from the past – in turn, considering each at different depths.

The control of plant modifications

According to the design drawings a six-inch (150 mm) diameter open hole was originally used as the tank vent. The plant had been in operation for about seven years and sometime during this period the six-inch hole had been blanked off and an unused three-inch (75 mm) opening on the tank used instead. (The tank was a second-hand one and was provided with more openings than were needed for its new duty.) No-one knew when the change had been made, who authorized it, or why. No records had been kept. It is possible that the six-inch hole allowed dirt to enter the tank and contaminate the product. Experience elsewhere on the plant, which had many similar tanks and handled many similar products, showed that while a three-inch hole could block with solid it was most unlikely that a six-inch one would do so.

Blanking the six-inch opening, which ended in a standard flange, and removing a blank from the three-inch opening, would have taken a fitter little more than an hour. The cost would have been negligible and no financial authorization would have been required. The only paperwork needed would have been a permit-to-work, which any of the shift foremen could have issued. Yet the modification led to the deaths of two men. The incident shows very clearly the results that can follow from unauthorized modifications, and the need for a system to control them. References 1–3 (and Chapters 1 and 8) describe other accidents that have resulted from modifications that had unforeseen side-effects, and the first two describe the essentials of a control system. This should consist of four parts:

(1) **Instructions.** No modification to plant or process should be made unless it has been authorized in writing by a competent person, who

should normally be professionally qualified. Before giving that authorization the competent person should try to identify all the consequences and should specify the change in detail. When the modification is complete, and before it is commissioned, he should inspect it to see that his intentions have been followed and that the modification 'looks right'.

(2) **Aids.** There should be some sort of guide sheet or check list to help people identify the consequences. One such guide sheet is shown in references 1 and 2.

(3) **Training.** People will only follow this procedure if they are convinced that it is necessary. A training programme, such as the one available from the Institution of Chemical Engineers[4], is needed.

(4) **Monitoring.** As with all safety procedures, regular auditing, to make sure that the system is being used, is necessary.

The vent was a relief valve

An open vent is in effect a relief vent. It is a very simple relief valve, but it fulfils the function of a relief valve and it should be treated with the same respect, i.e. it should be registered for regular inspection (to make sure that its size has not been changed and that it is not obstructed) and it should not be altered in any way without going through the same procedure as we go through before modifying a relief valve.

Unfortunately those concerned did not realize this. No process manager, foreman or operator would dream of asking a fitter to replace a relief valve by a smaller one unless calculation has shown that a smaller one is adequate, and no fitter would carry out the replacement. Yet replacement of a six-inch vent by a three-inch one seems to have been done at the drop of a hat. There are still many people who treat open vents with less respect than relief valves. Let us hope they learn from this incident and do not wait until a similar one occurs on their own plant.

As stated in Chapter 1, equipment whose size or existence has been taken into account in sizing a relief valve, such as non-return (check) valves, control valve trims and restriction orifice plates, should also be registered for regular inspection and should not be modified or removed without checking that the size of the relief valve will still be adequate.

As already stated, the six-inch hole was fitted with a standard flange, a survival from the vessel's previous duty. This made it too easy for the vent to be blanked. Vent openings should be designed so that they cannot easily be blanked.

The choking of the vent

Choking of the vent could have been prevented by retaining its original size. It could also have been prevented by heating the vent (with steam or electric tracing) and/or by inspecting the vent to see if it was choked and rodding it clear if necessary. Some operators did this but the access was poor and others did not bother. After the accident it came to light that one of the shift foremen had complained to the manager several times about the difficult access and had asked for improvements to be made but nothing had been done.

Nothing had been done because the choking of the vent was looked upon, by all concerned, as an inconvenience rather than a hazard, and improvements got the priority given to an inconvenience, not that given to a hazard. Nobody realized that the vessel was too weak to withstand the air pressure, if the vent was fully choked.

After the accident, changes were made to the vent, and to the vents on 200 similar tanks. The vents were steam-jacketed, fitted with flame traps (the reason for these is described later) and, in case the vents should choke, a steam-heated blow-off panel was provided (see page 59 for details). The access was also improved. Each vent was inspected every day to make sure that it was clear and that the steam heating was in operation. The frequency of inspection was much higher than for a normal relief valve, where once every two years is considered adequate, but was necessary as the vents can choke quickly if the steam supply fails. A special inspector had to be appointed, as the volume of work involved was considered too much for the operators.

There may be many other operators, and managers, who do not know the strength of their equipment. All plant staff should know the pressures, temperatures, concentrations of corrosive materials etc. that their equipment can safely withstand and the results that will follow if they go beyond them.

The operators found it hard to believe that a 'puff of air' had blown the end of the tank, and explosion experts had to be brought in to convince them that a chemical explosion had not occurred. On many other occasions operators have been quite unaware of the power of compressed gases[5].

Mitigation of consequences

An ordinary atmospheric-pressure storage tank would have been adequate for the storage of the product. A second-hand pressure vessel was used, as it happened to be available. The designer probably thought that by using a stronger vessel than necessary he was providing increased safety. In fact he was not. If an atmospheric-pressure tank had been used and its vent had choked, the roof would have blown off at a gauge pressure of about 24 inches of water (less than $1\,lbf/in^2$ or about $6\,kPa$), not $20\,lbf/in^2$, and the energy released would have been correspondingly less, though still high enough to kill someone who was standing in the direction of failure. However, as atmospheric-pressure storage tanks are vertical cylinders, normally with weak seam roofs which blow off if the tank is overpressured, it is unlikely that anyone will be in the way. In contrast, if the end comes off a horizontal cylinder, someone is more likely to be in the way.

Similarly, if vessels are exposed to fire, weaker vessels fail at lower pressures, with less disastrous consequences, than stronger vessels[6].

Designers should therefore:

• Remember that stronger does not always mean safer.
• Consider the way in which equipment will fail.

Alternatively, the tank could have been designed to withstand the full pressure of the compressed air, but this would be an expensive solution.

Could it have been a chemical explosion?

The product in the tank had a flashpoint of 120°C and was kept at about 150°C, so a flammable atmosphere was present in the tank. There was no evidence that a chemical explosion had occurred and the tank failed at exactly the time that it was calculated the pressure would have risen to the bursting pressure. Nevertheless, it is not good practice to blow compressed air into a tank containing flammable vapour, and after the incident, nitrogen was used instead. As air might diffuse into the tank, a flame trap was fitted in the vent. Two hundred other tanks in the plant were treated similarly.

The flame trap assembly[8]

Figures 7.2 and 7.3 illustrate the flame trap assembly that was fitted to 200 tanks, in place of the manhole cover, after the accident. The device consists of a short length of steam-jacketed pipe into which a crimped ribbon flame arrester is fitted. A long handle on the flame arrester enables it to be removed, without the use of tools, and held up to the light for inspection. The inspector can also look through the vent hole into the tank to see that there is no obstruction. Adequate access must, of course, be provided. Should the flame arrester choke, or be unable to pass the required quantity of gas, a loose D-shaped lid lifts and relieves the pressure. A chain prevents the lid going into orbit. Steam tracing prevents the lid from being stuck in position by adhering solid.

The steam jacket should extend at least 3 inches (75 mm) into the tank to prevent a bridge of solid forming across the bottom. The flame arrester should be about 9 inches (225 mm) from the bottom of the vent and about 3 or 4 inches (75 or 100 mm) from the top.

Manufacturers of flame arresters can supply data on the gas flows through them at various pressure drops. For example, a 4.66 inch (118 mm) diameter flame arrester, ¾ inch (20 mm) thick, will pass 700 m³/hr of air at a pressure drop of 8 inches (200 mm) of water (2 kPa).

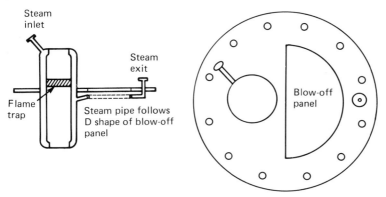

Figure 7.2 The flame trap assembly which was fitted in place of the man-hole cover. (a) Elevation section. (b) Plan from above

Figure 7.3 The flame trap assembly, showing how the flame trap can be removed for inspection

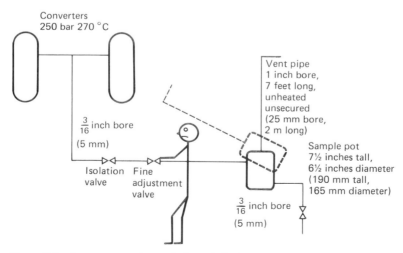

Figure 7.4 An attempt to clear a choke and obtain a sample from the line joining two reactors

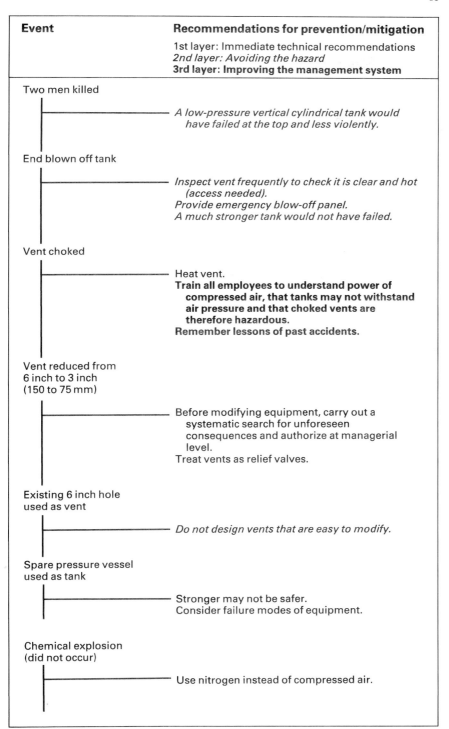

Event	Recommendations for prevention/mitigation
	1st layer: Immediate technical recommendations *2nd layer: Avoiding the hazard* **3rd layer: Improving the management system**

Two men killed

— *A low-pressure vertical cylindrical tank would have failed at the top and less violently.*

End blown off tank

— Inspect vent frequently to check it is clear and hot (access needed).
Provide emergency blow-off panel.
A much stronger tank would not have failed.

Vent choked

— Heat vent.
Train all employees to understand power of compressed air, that tanks may not withstand air pressure and that choked vents are therefore hazardous.
Remember lessons of past accidents.

Vent reduced from
6 inch to 3 inch
(150 to 75 mm)

— Before modifying equipment, carry out a systematic search for unforeseen consequences and authorize at managerial level.
Treat vents as relief valves.

Existing 6 inch hole
used as vent

— *Do not design vents that are easy to modify.*

Spare pressure vessel
used as tank

— Stronger may not be safer.
Consider failure modes of equipment.

Chemical explosion
(did not occur)

— Use nitrogen instead of compressed air.

Figure 7.5 Summary of Chapter 7

Learning from the past

During the investigation someone recalled that a similar accident had occurred about 20 years before, and after much searching the report was found. Another man had been killed by a choked vent about 100 metres away, although in a plant under the control of a different department.

A vegetable oil was being hydrogenated on a pilot plant and samples had to be taken from the line joining two reactors (Figure 7.4). The sample pot was filled by opening the isolation valve and cracking open the fine-adjustment valve. Both valves were then closed and the sample drained into a sample can. As the oil and product had a melting point of 20°C, all lines were steam-traced, except for the vent line.

Two weeks after start-up the operator was unable to get a sample. Suspecting a choke, his chargehand tried to clear it by opening the isolation and drain valves fully and then gradually opening the fine-adjustment valve. The vent pipe moved suddenly and hit the chargehand on the head.

The sample pot and vent line had not been clamped, and could be made free to rotate by partially unscrewing the flange on the inlet pipe. It had been intended to clamp them but this had been overlooked. It is believed that there was a choke somewhere in the system, probably in the vent pipe, which was not heated, and that when it cleared the back-pressure from the sudden rush of gas caused the vent pipe to move backwards.

The main recommendation of the report was that plants handling materials which are solid at atmospheric temperature should be adequately heated. This applied particularly to the vent pipe.

Many other examples could be quoted of serious accidents which have been forgotten, or not made known to other departments of the same organization, and which have been repeated after 10 or more years[7]. Another was described in Chapter 4. (Less serious accidents are forgotten and repeated more often[8].) This will continue unless companies take action to keep memories alive and remind people of the lessons of past accidents. This is discussed further in Chapter 17.

References

Brief accounts of this accident originally appeared in references 7 and 9.

1. KLETZ, T. A. *Chemical Engineering Progress,* Vol. 72, No. 11, Nov. 1976, p. 48.
2. LEES, F. P. *Loss Prevention in the Process Industries,* Butterworths, 1980, Vol. 2, Sections 21.6–21.8.
3. KLETZ, T. A. *What went wrong? Case Histories of Process Plant Disasters,* Gulf Publishing Co., 1985, Chapter 2.
4. *Hazards of Plant Modifications – Hazard Workshop Module No. 002,* Institution of Chemical Engineers, Rugby, UK, 1978.
5. KLETZ, T. A. *An Engineer's View of Human Error,* Institution of Chemical Engineers, Rugby, UK, 1985, p. 7.
6. KLETZ, T. A. *Myths of the Chemical Industry,* Institution of Chemical Engineers, Rugby, UK, 1984, p. 14.
7. KLETZ, T. A. *Loss Prevention,* Vol. 13, 1980, p. 1.
8. KLETZ, T. A. *Loss Prevention,* Vol. 10, 1976, p. 151.
9. KLETZ, T. A. *Plant/Operations Progress,* Vol. 1, No. 4, Oct. 1982, p. 252.

Chapter 8

Flixborough

The explosion in the Nypro factory at Flixborough, England, on 1 June 1974 was a milestone in the history of the chemical industry in the UK. The destruction of the plant in one almighty explosion, the death of 28 men on site, and extensive damage and injuries, though no deaths, in the surrounding villages, showed that the hazards of the chemical industry were greater than had been generally believed by the public at large. In response to public concern the government set up not only an inquiry into the immediate causes but also an 'Advisory Committee on Major Hazards' to consider the wider questions raised by the explosion. Their three reports[1] led to far-reaching changes in the procedures for the control of major industrial hazards. The long-term effects of the explosion thus extended far beyond the factory fence.

The plant on which the explosion occurred was part of a complex for the manufacture of nylon, jointly owned by Dutch State Mines (55%) and the UK National Coal Board (45%). It oxidized cyclohexane, a hydrocarbon similar to petrol in its physical properties, with air to a mixture of cyclohexanone and cyclohexanol, usually known as KA (ketone/aldehyde) mixture. The reaction was slow and the conversion had to be kept low to avoid the production of unwanted byproducts, so the inventory in the plant was large, many hundreds of tonnes. The reaction took place in the liquid phase in six reaction vessels, each holding about 20 tonnes. The liquid overflowed from one reactor to another while fresh air was added to each reactor (Figure 8.1). Unconverted raw material was recovered in a distillation section and recycled. Similar processes were operated, with variations in detail, in many plants throughout the world.

One of the reactors developed a crack and was removed for repair. In order to maintain production, a temporary pipe was installed in its place. Because the reactors were mounted on a sort of staircase, so that liquid would overflow from one to another, this temporary pipe was not straight but contained two bends (Figure 8.1). It was 20 inches (500 mm) in diameter, although the short pipes which normally joined the reactors together were 28 inches (700 mm) in diameter. Calculation showed that 20 inches would be adequate for the flow rates required. Bellows, also 28 inches in diameter, were installed between each reactor and these were left at each end of the temporary pipe.

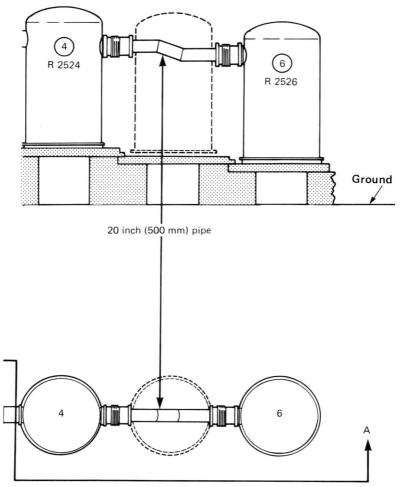

Figure 8.1 Arrangement of reactors and temporary pipe at Flixborough. (Reproduced by permission of the Controller of Her Majesty's Stationery Office)

The men who were charged with the task of making and installing the temporary pipe had great practical experience and drive – they constructed and installed it and had the plant back on-line in a few days. However, they were not professionally qualified and did not realize that the pipe and its supports should have been designed by an expert in piping design. Their only drawing was a full-size sketch in chalk on the workshop floor. The only support was a scaffolding structure on which the pipe rested.

The temporary pipe performed satisfactorily for two months until a slight rise in pressure occurred. The pressure was still well below the relief valve set-point, but nevertheless it caused the temporary pipe to twist (Figure 8.2). The bending moment was strong enough to tear the bellows and two 28 inch holes appeared in the plant. The cyclohexane was at a

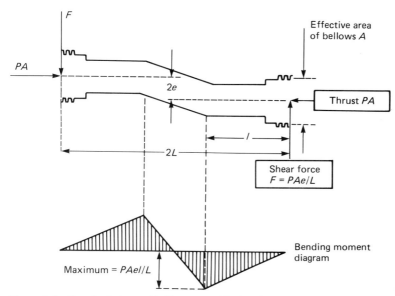

Figure 8.2 Sketch of pipe and bellows assembly showing shear forces on the bellows and bending moments in the pipe (due to internal pressure only). (Reproduced by permission of the Controller of Her Majesty's Stationery Office)

gauge pressure of about 10 bar (150 lbf/in^2) and a temperature of about 150°C. It was thus under pressure above its normal boiling point (81°C) and a massive leak occurred as the pressure was reduced. It was estimated that about 30–50 tonnes escaped in the 50 seconds that elapsed before ignition occurred. The source of ignition was probably a furnace some distance away.

The resulting explosion, one of the worst vapour cloud explosions that has ever occurred, destroyed the oxidation unit and neighbouring units and caused extensive damage to the rest of the site. In particular the company office block, about 100 metres away, was destroyed, and had the explosion occurred during office hours, rather than at 5 p.m. on a Saturday, the death toll might have been 128 instead of 28. There is a detailed account of the explosion and of the events leading up to it in the official report[2] and in reference 3.

Instead of looking at the outer layers of the onion first and then the inner layers, it will be convenient to look at the various events leading up to the accident, summarized in Figure 8.4, in the order in which they occurred.

Reduction of inventory

One of the most important lessons to be learned from Flixborough, perhaps the most important, was ignored in the official report and in most of the other papers which have appeared on the subject: *If the inventory had been smaller the leak would have been less. What you don't have, can't leak.* In fact, it was Flixborough that made chemical engineers realize that it might be possible to reduce inventories. Before Flixborough most companies designed plants and accepted whatever inventory was called for by the design. Many companies still do so. However, an increasing number now realize that if we set out to reduce inventories it is often possible to do so and that the resulting plants are cheaper as well as safer. They are cheaper because less added-on protective equipment is needed, and has to be tested and maintained, and because a smaller inventory means smaller, and therefore cheaper, equipment. The author's book *Cheaper, Safer Plants*[4] gives many examples of what has been and might be done. Such plants are said to be inherently safer, to distinguish them from conventional or extrinsically safe plants, where the safety is obtained by adding on safety features.

Let us look at the scope for inventory reduction in the process operated at Flixborough. The output of the plant was about 50 000 tonnes/yr KA. Assuming a linear velocity of 1 m/s, it could all have passed through a pipe 1.6 inches (41 mm) in diameter. Any larger pipe is a measure of the inefficiency of the chemical engineering. Actual pipe sizes ranged up to 28 inches (700 mm), so the cross-section of the pipe, and thus the flow rate, was $(28/1.6)^2 = 300$ times greater than the theoretical minimum. This is a bit unfair, as the 28 inch pipes were made larger than necessary for the liquid flow so that the gas pressures in the reactors were the same. Photographs of the plant suggest that liquid lines were at least 12 inches (300 mm) in diameter. The 'inefficiency' or 'hazard enhancement' index was thus $(12/1.6)^2 = 56$. Was it possible to reduce it?

The inefficiency was so large because reaction was slow and conversion low (5–10%). The reaction takes place in two stages, oxidation to a peroxide followed by its decomposition. The first stage is not really slow. Once the molecules come together they react quickly. It is the mixing that is slow, the chemical engineering rather than the chemistry, so if the mixing can be improved, reaction volume can be reduced. The second stage is slow and a long residence time is therefore required. If it could take place in a long thin tube instead of a vessel, leaks would be unlikely, would be restricted by the size of the tube and could be stopped by closing an emergency valve. Higher temperatures would increase the rate of decomposition. Conversion was low because if more air is used to increase conversion it is difficult to ensure that the air is intimately mixed with the hydrocarbon, and high local concentrations cause unwanted byproducts to be formed. If a better method of mixing air and hydrocarbon can be found then conversion can be increased.

Another problem that would have to be overcome in a new design would be cooling. The heat of reaction is considerable and is removed by evaporation and condensation of unconverted cyclohexane, and its return to the reactor. About 7 tonnes have to be evaporated for every tonne that reacts. A higher conversion process would require a large heat-exchange area, a property of long thin tubes, unless some other method of energy removal, such as electricity generation in fuel cells, can be devised[5].

Of course, the company concerned, Nypro, cannot be blamed for adopting a high-inventory process. They were using much the same process as every other manufacturer of nylon and were following 'best current practice'. It simply had not occurred to the industry as a whole that inventory reduction was either possible or desirable. The chance of a major escape was considered so low that the large inventory was not looked upon as a hazard and everyone was so used to the process that its gross inefficiency was accepted without comment or complaint. Yet most of the material in the plant was getting a free ride; it was going round and round with only a small proportion being reacted and removed at each pass. What would we think of an airline which flew from London to New York and back with only 5% of the passengers alighting on each round trip, the rest staying on board to enjoy the movies?

Following Flixborough one company did devise a lower-inventory process but abandoned the development programme when it became clear that no new plants would be needed in the foreseeable future. More recently Union Carbide have described an improved pot reactor for gas/liquid reactions[6]. The gas is added to the vapour phase and is pulled down into the liquid by a down-pumping impeller which is used in place of a conventional stirrer. Higher gas solution rates are said to be obtainable without the risk of high local concentrations, and the conversion is improved.

The Flixborough explosion thus demonstrates the desirability of reducing inventories. It also shows that the most important inventories to reduce are those of flashing flammable or toxic liquids, i.e. liquids under pressure above their normal boiling points. Suppose a 2 inch (50 mm) hole appears in a line or vessel carrying a liquid such as petrol at a gauge pressure of 7 bar and at atmospheric temperature. Liquid will leak out at a rate of about 3 tonnes/min but very little vapour will be formed (unless the liquid leaks at a high level and falls over the structure) and the chance of an explosion or large toxic cloud is very small. If the line or vessel contains a gas such as propane at the same pressure and temperature, the leak will be much smaller, about ¼ tonne/min, it may disperse by jet mixing and again the vapour cloud will not be large. However, if the line or vessel contains petrol at the same pressure but at a temperature of 100°C, above its normal boiling point, the liquid will again leak at a rate of about 3 tonnes/min but will then turn to vapour and spray. The spray is just as explosive and just as toxic as the vapour and a large explosive or toxic cloud will be formed. Most unconfined vapour cloud explosions have resulted from leaks of flashing flammable liquids such as liquefied petroleum gas or hot hydrocarbon under pressure[7], and most toxic incidents from leaks of liquefied toxic gases such as chlorine or ammonia.

Since 1974 progress towards inherently safer plants has been slower than was hoped. In part this has been due to the recession in the chemical industry – few new plants have been built – but there are also other reasons: our natural caution and conservatism when innovation is involved and the fact that inherently safer designs require more time during the early stages of design for alternatives to be evaluated and new ideas developed, time which is often not available if the plant is to be ready in time to meet a market opportunity. This applies to all innovation but particularly to innovation in the field of safety, as in most companies safety studies do not take place and safety people do not get involved until late in design. If we are to develop inherently safer designs, we need to carry out studies similar to hazard and operability studies (hazops) much earlier in design, at the conceptual stage, when we are deciding which process to use and at the flowsheet stage, before detailed design starts. These studies should be additional to the usual hazop of the line diagram and will not replace it. In addition, Malpas[8] suggests that during design one member of the design team should be responsible for thinking about the plant after next, as during design we are conscious of all the changes we would like to make but cannot make as there is insufficient time. These constraints on the development of inherently safer designs are discussed in more detail in reference 4.

The control of process modifications

Why did a crack appear in one of the reactors? There was a leak of cyclohexane from the stirrer gland and, to condense the leaking vapour, water was poured over the top of the reactor. Plant cooling water was used, as it was conveniently available.

Unfortunately the water contained nitrates and they caused stress corrosion cracking of the mild steel pressure vessel (though fortunately not of the stainless steel liner). Nitrate-induced cracking was well-known to metallurgists but was not well-known at the time to other engineers[9].

The next section of this chapter and the explosion described in Chapter 7 show the importance of controlling plant modifications. Process modifications, as this crack shows, can also produce unforeseen results. No change in operating conditions, outside the approved range, should be made without going through the procedure described in Chapter 7, i.e. they should not be made until they have been authorized by a professionally qualified manager, who should first use a systematic technique to help him identify the consequences. Pouring water over equipment was at one time a common way of providing extra cooling but it is nevertheless a process change that should go through the modification procedure. *The more innocuous a modification appears to be, the further its influence will extend.*

However, no great blame can be attached to the Nypro staff. They did what many people would have done at the time. The purpose of investigations, and of this book, is not to find culprits but to prevent future accidents.

The control of plant modifications

As already described, the temporary pipe was installed very quickly, to restore production, and there was no systematic consideration of the hazards or consequences. One of the main lessons to be learned is that stated in Chapter 7, that no modification should be made until it has been authorized in writing by a competent person who should normally be professionally qualified, and that before giving that authorization the competent person should try to identify all the consequences and should specify the change in detail. When the modification is complete and before it is commissioned he should inspect it to see that his intentions have been followed and that the modification 'looks right'. In particular, the modification should be made to the same standard as the original design and this standard should be specified in detail. In addition there should be a guide sheet or check list to help the competent person identify the consequences, and a training programme is needed to convince people that the modification control procedure is necessary. As with all procedures, regular monitoring is needed to make sure that the procedure is being followed.

Note that modifications to supports can be as important as modifications to equipment and that if pipes are being modified the supports need as much consideration as the pipe itself.

The need for suitably qualified staff

The reference to professional qualifications is important. The men who constructed the temporary pipe did not know how to design large pipes capable of withstanding high temperatures and pressures. Few engineers do. It is a specialized branch of mechanical engineering. However, a professional engineer would have recognized the need to call in an expert in piping design. The men who constructed the pipe did not even know that expert knowledge was needed: *they did not know what they did not know.* As a result they produced a pipe that was quite incapable of withstanding the operating conditions. In particular, to install an unrestrained pipe between two bellows was a major error, specifically forbidden in the bellows manufacturer's literature.

At the time the pipe was constructed and installed there was no professionally qualified mechanical engineer on site, though there were many chemical engineers. The establishment called for one professional mechanical engineer, the works engineer, but he had left and his successor, though appointed, had not yet arrived. Arrangements had been made for a senior engineer of the National Coal Board, who owned 45% of the plant, to be available for consultation, but the men who built the pipe did not see the need to consult him. Another lesson of Flixborough is, therefore, the need to see that the plant staff are a balanced team, containing people of the necessary professional experience and expertise. On a chemical plant mechanical engineering is as important as chemistry or chemical engineering. To quote a former chief engineer of ICI's Billingham Division, 'A place like Billingham is really engineering; chemistry is only what goes through the pipes'[10].

Since Flixborough, most chemical plants have reduced staff but the need for professional expertise remains as great as ever. For example, a plant which at one time employed an electrical engineer may no longer do so, the control engineer having to act as electrical engineer as well. There is an electrical engineer available for consultation somewhere in the company, but will the control engineer know when to consult him? Will he know what he does not know?

Should the chemical engineers have noticed that the temporary pipe did not look right? At the time some of my colleagues said that if they had been on the plant they would have questioned its design. Others thought that it was unreasonable to expect a chemical engineer to do so, and certainly those who were on the plant cannot be blamed for not noticing mechanical design errors. The explosion does, nevertheless, show the importance of plant staff, at all levels from the most senior downwards, spending part of each day out on the plant looking round. A plant cannot be managed from an office. Unfortunately too many managers find that the volume of paperwork provides a good excuse for staying in the warmth and comfort of their offices.

If a chemical engineer had felt that the appearance of the temporary pipe was not right he might have felt that it was none of his business. Perhaps the mechanical engineers would resent his interference. Presumably they know their job. Better say nothing. Flixborough shows us that we should never hesitate to speak out if we suspect that something may be wrong.

In the light of Flixborough, companies should ask themselves not only if they have enough mechanical engineers, electrical engineers etc. but also if their safety advisers have the right technical knowledge and experience and sufficient status for their opinions to carry weight. In high-technology industries the safety adviser needs to be qualified and experienced in the technology of the industry, with previous line management experience, or the managers will not listen to him. He needs to be numerate, able to assess risks systematically and numerically, and a good communicator[11].

Robust equipment preferred

No criticism can be made of the manufacturers of the bellows, which were misused in a way specifically forbidden in their literature. Nevertheless bellows are inevitably a weak link, they will not withstand poor installation to the same extent as fixed pipework, and they should not be used when hazardous materials are handled. Instead, expansion loops should be built into the pipework.

Generalizing, we should, whenever possible, avoid equipment which will not withstand mistreatment. Flexible hoses are another example. To use a computer phrase, whenever possible we should use equipment which is 'user-friendly'.

Plant layout and location

It is almost impossible to prevent ignition of a leak the size of that which occurred at Flixborough. It will spread until it reaches a source of ignition. However, it is possible to locate and lay out a plant so that injuries and damage are minimized if an explosion occurs.

Most of the men who died were in the control room and were killed by the collapse of the building on them. Since Flixborough many companies have built blast-resistant control rooms, able to withstand a peak reflected pressure of 0.7 bar ($10 \, lbf/in^2$). They are not designed to withstand repeated explosions, but to withstand one explosion. They may be damaged but should not collapse, thus protecting the men inside and also preserving the plant records which may help the subsequent investigation. Reference 12 gives a design procedure.

Is it right to protect the people in the control room and ignore those who are outside? At Flixborough most of those outside survived. People in an ordinary unreinforced building are at greater risk than those outside.

The control room has to be near the plant but other occupied buildings can usually be placed further away so that they need not be strengthened. Office blocks and other buildings containing large numbers of people should certainly never be located close to the plant, as at Flixborough. If, for any reason, an occupied building has to be near the plant, then it should be strengthened. How near is near? The following procedure is suggested in reference 13:

(1) Estimate the largest leak that should be considered, ignoring vessel failures, taking pipe failures into account, and also allowing for remotely operated emergency isolation valves.
(2) Estimate the size of the vapour cloud, by doubling the theoretical flash to allow for spray.
(3) Estimate the pressure developed at various distances if this cloud ignites. It is necessary to allow for the efficiency of the explosion, and a figure of 2% is suggested. However, some writers have suggested higher figures.
(4) Design the building to withstand the pressure developed at the point at which it is placed.

The same procedure can be used to find a satisfactory layout and location for the plant. Thus other plants containing hazardous materials should not be placed at points at which the peak incident pressure will be 0.35 bar (5 lbf/in²) or more, or they may be so seriously damaged that a further leak and a further explosion occur (the so-called domino effect). For low-pressure storage tanks containing hazardous materials the maximum acceptable pressure is 0.2 bar (3 lbf/in²). Public roads should not be located where the pressure exceeds 0.07 bar (1 lbf/in²) and houses where it exceeds 0.05 bar (0.7 lbf/in²), preferably 0.03 bar (0.4 lbf/in²). The conclusions are summarized in Figure 8.3 and more details are given in reference 12.

Figure 8.3 shows that there should be no buildings at all within 20 metres of plant containing materials which might explode. This is to provide a fire barrier. It is good practice to divide large plants into blocks, separated by gaps about 15–20 metres wide.

Some writers have suggested lower pressures, some higher. However, they have also suggested different methods for the calculation of the pressure, in particular different explosion efficiencies, and their final conclusions are not much different from those shown in Figure 8.3. Each method of estimation should be considered as a whole and we should not pick and choose figures from different methods.

A few companies have questioned the need for the precautions described above. It would be better, they say, to prevent explosions than guard against the consequences. This is true and we do try to prevent explosions, but at the present time experience shows that the probability of an explosion is not so low that it can be ignored.

A final point to be considered under layout is the location of the assembly points at which people congregate when a warning sounds and a

72

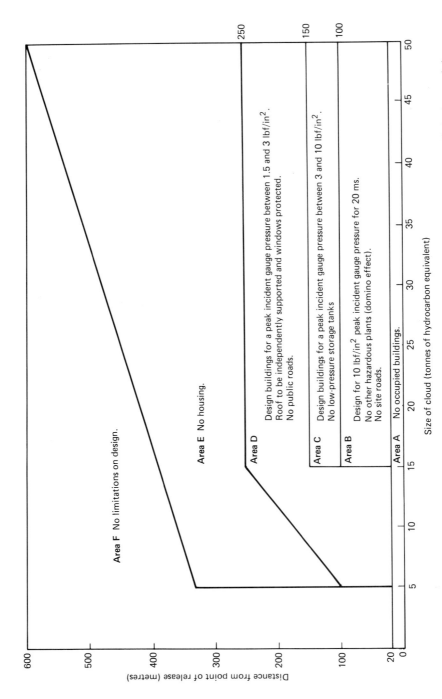

Figure 8.3 Summary of the main restrictions on design imposed by unconfirmed vapour cloud explosions. (Note: Area A limitations apply in areas E–A and so on)

plant is evacuated. They should be some distance away, at least 150 metres, from equipment containing large inventories, in Area D of Figure 8.3. Control rooms should not be used as assembly points.

Layout is considered in more detail in reference 14.

Prevention of leaks

The Flixborough explosion was followed by an explosion of papers on the probability of large leaks and on their behaviour: how far will they spread in various wind and weather conditions, what pressures will be developed if they ignite? etc. In contrast, there has been very little written on the reasons for other large leaks and the actions needed to prevent them. It is as if the occasional large leak is considered inevitable.

It will be shown in Chapter 16 that about half the large leaks that have occurred have been the result of pipe failures and that many pipe failures have occurred because construction teams did not follow the design in detail or did not do well what was left to their discretion. The most effective single action we can take to prevent major leaks is therefore to specify piping designs in detail and then to check during and after construction that the design has been followed and that items not specified in the design have been constructed in accordance with good engineering practice. Much more thorough checking is needed than has been customary in the past.

Public response

The material damage at Flixborough cost the insurance companies about £50M (equivalent to well over £200M [$300M] at 1987 prices). It cost the industry many times more in the extra equipment they were required to install, and the extra procedures they were required to carry out. Much of this expenditure should have been undertaken in any case. Some of it went further than was really necessary and was the result of the understandable public reaction to the explosion. It demonstrated that in the eyes of the public the industry is one and that the industry as a whole is held responsible for, and suffers from, the shortcomings of individual companies. It therefore shows that companies should freely exchange information on accidents that have happened and the action needed to prevent them happening again. Unfortunately companies are now less willing to exchange information than they were in the 1970s. The reasons are discussed in detail in reference 15. The most important reasons are time – with reductions in numbers managers have less time to do the things they know they ought to do but do not have to do by a specific date – and, particularly in the US, the influence of company lawyers, who fear that disclosure may influence claims for damages or prosecution.

In the UK the government, as already stated, set up an Advisory Committee on Major Hazards to consider the wider implications of the Flixborough explosion. It took about 10 years for their recommendations to be made and to come into force and they resulted in the CIMAH

(Control of Industrial Major Accident Hazards) Regulations[16]. These require companies which have more than defined quantities of hazardous materials in storage or being processed to prepare a 'safety case' describing the action they have taken to identify and control the hazards. Two years after Flixborough the Seveso accident (Chapter 9) caused a similar public reaction across Europe and resulted in the so-called Seveso Directive of the European Community[17]. The CIMAH Regulations, as well as being the UK's response to Flixborough, also brought the Sevesco Directive into force in the UK. In the US there was no similar response until after Bhopal (Chapter 10).

The CIMAH Regulations were made under the Health and Safety at Work Act and like the Act are 'inductive', i.e. they state an objective that has to be achieved but do not say how it must be achieved, though advice is available. In some countries, including the US, it seems to be believed that industrial accidents can be prevented by the government writing a book of regulations that looks like a telephone directory, though it is rather less interesting to read. In other countries, including the UK, it has been realized that it is impracticable to control complex, developing, high-technology industries by detailed regulations. They soon get out-of-date or circumstances arise which were not thought of by those who wrote them, and they encourage the belief that all you need to do to get a safe plant is to follow the rules. Under the UK Health and Safety at Work Act, passed by Parliament at the time of Flixborough but not then fully in force, there is a general obligation to provide a safe plant and system of work and adequate instruction, training and supervision, so far as is reasonably practicable. In the first place it is up to the employer to decide what is a safe plant etc., but if the Health and Safety Executive (HSE) – the body responsible for enforcing the Act – do not agree they will say so and if necessary issue an improvement or prohibition notice.

The UK approach is summed up in the following extract from an article entitled *Can HSE Prevent another Flixborough?*, written by a member of the HSE:[18] 'On his own, the inspector is unlikely ever to be able to provide adequate safeguards or deterrents. He must and should rely heavily on industry where the expertise in a particular field is surely to be found. He must however learn the know-how necessary to identify and follow up weaknesses in both management and systems'.

Nevertheless, many people in the chemical industry feel that the public and governmental reaction to Flixborough has been excessive; they contrast it with the relative indifference to some other industrial hazards, such as those of the construction industry, and non-industrial hazards, such as the roads. Perhaps the industry needs to do more to explain the true size of the hazards and, equally important, the benefit the public get back in return. The public accept a terrible death toll on the roads because the advantages of the motor car are clear and obvious to all. In contrast, the chemical industry is seen as making unpleasant chemicals with unpro-nounceable names in order to increase its sordid profits. At the best it provides exports and employment. The public does not realize that it provides the necessities for our modern way of life.

The Flixborough explosion affected the insurance companies as well as the public. It made many of them realize that individual claims might be larger than they had previously foreseen and many of them increased their premiums. It also made them ask themselves if they were doing enough to distinguish between good and bad risks, and many of them increased their investment in systematic methods of risk assessment, thus accelerating a trend that was already under way. It is, of course, relatively easy for them to recognize those plants which are better designed. It is much more difficult to distinguish between good and bad management.

The responsibilities of partnerships

The Flixborough plant was jointly owned by Dutch State Mines, who supplied the know-how, and the UK National Coal Board. In such joint ventures it is important to be clear who is responsible for safety in both design and operation. Shortcomings in this respect do not seem to have been responsible in any way for the Flixborough explosion but they were relevant at Bhopal (Chapter 10).

The rebuilt plant

When the plant was rebuilt, the product, cyclohexanol, was manufactured by a safer route, much to the relief of the local population. Instead of manufacturing it by the oxidation of cyclohexane, it was made by the hydrogenation of phenol. It is doubtful if the company would have been allowed to rebuild the plant if they had wished to use the original process.

However, phenol itself has to be manufactured, usually by the oxidation of cumene to cumene hydroperoxide and its 'cleavage' to phenol and acetone. This process, judged by its record, is at least as hazardous, perhaps more hazardous, than the oxidation of cyclohexane. It was not carried out at Flixborough, but elsewhere. The hazards were not really reduced, only exported.

The rebuilt plant had a short life. It was closed down, after a few years, for commercial reasons.

Progress in implementing recommendations

This has been a long chapter, so in conclusion it may be useful to summarize the progress made in carrying out the various recommendations made after Flixborough.

(1) On some issues, progress has been good. Much more attention is now paid to plant layout and location and the strengthening of control buildings, and guidance is available for the designer. Many companies have established procedures for the control of modifications. Insurance companies play a more active role in encouraging good design. The competence of safety advisers has improved.

(2) In contrast, progress in the development of inherently safer designs has been disappointing. Little has been done to develop new designs and full use is not being made of the designs that are available. Little attention has been paid to the reasons for leaks and ways of preventing them.

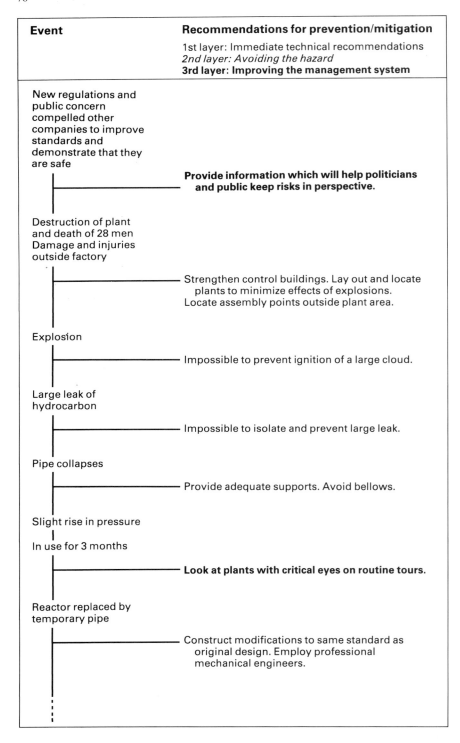

Event	Recommendations for prevention/mitigation
	1st layer: Immediate technical recommendations *2nd layer: Avoiding the hazard* **3rd layer: Improving the management system**
New regulations and public concern compelled other companies to improve standards and demonstrate that they are safe	
	Provide information which will help politicians and public keep risks in perspective.
Destruction of plant and death of 28 men Damage and injuries outside factory	
	Strengthen control buildings. Lay out and locate plants to minimize effects of explosions. Locate assembly points outside plant area.
Explosion	
	Impossible to prevent ignition of a large cloud.
Large leak of hydrocarbon	
	Impossible to isolate and prevent large leak.
Pipe collapses	
	Provide adequate supports. Avoid bellows.
Slight rise in pressure	
In use for 3 months	
	Look at plants with critical eyes on routine tours.
Reactor replaced by temporary pipe	
	Construct modifications to same standard as original design. Employ professional mechanical engineers.

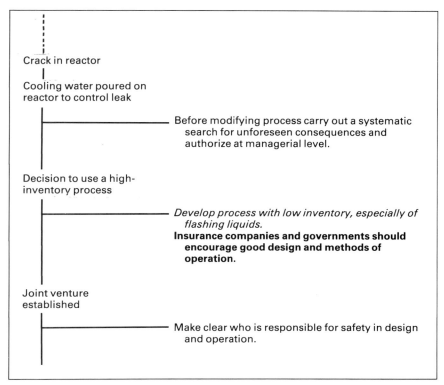

Figure 8.4   Summary of Chapter 8

Progress in these areas is important because while we can control small leaks by the methods discussed in Chapter 4 – emergency isolation valves, steam and water curtains, open construction etc. – we cannot control leaks of Flixborough size. Once a leak of this size occurs, all we can do is hope that it will not ignite (and, if time permits, evacuate the plant).

(3) In other areas progress varies greatly from company to company or even between different plants belonging to the same company. Many managers spend much of their time out on the plant while others do not.

(4) In some countries governments are seen as playing a helpful role, at least by the more safety conscious sections of industry. In others it is questioned whether their activities result in a net increase in safety. The public generally have a poor understanding of the chemical industry's contribution to our way of life and of its safety record.

References

1. Advisory Committee on Major Hazards, *First, Second and Third Reports,* Her Majesty's Stationery Office, London, 1976, 1979 and 1984.

2. PARKER, R. J. (Chairman), *The Flixborough Disaster: Report of the Court of Inquiry,* Her Majesty's Stationery Office, London, 1975.
3. LEES, F. P. *Loss Prevention in the Process Industries,* Butterworths, 1980, Vol. 2, Appendix 1.
4. KLETZ, T. A. *Cheaper, Safer Plants,* Institution of Chemical Engineers, Rugby, UK, 2nd edition, 1985.
5. RAMSHAW, C. Private communication.
6. LITZ, L. M. *Chemical Engineering Progress,* Vol. 81, No. 11, Nov. 1985, p. 30.
7. KLETZ, T. A. *The Chemical Engineer,* No. 426, June 1986, p. 62.
8. MALPAS, R. in *Research and Innovation for the 1990's,* edited by B. Atkinson, Institution of Chemical Engineers, Rugby, UK, 1986, p. 26.
9. *Guide Notes on the Safe Use of Stainless Steel in Chemical Process Plant,* Institution of Chemical Engineers, Rugby, UK, 1978.
10. MAYNE, P. about 1950, quoted in *Chemical Age,* 20 Sep. 1974, p. 21.
11. KLETZ, T. A. *Safety and Loss Prevention,* Symposium Series No. 73, Institution of Chemical Engineers, Rugby, UK, 1982, p. B1.
12. *An Approach to the Categorization of Process Plant Hazards and Control Building Design,* Chemical Industries Association, 1979.
13. KLETZ, T. A. *Loss Prevention,* Vol. 13, 1980, p. 147.
14. MECKLENBURGH, J. C. (ed.), *Process Plant Layout,* Godwin, London, 1985.
15. KLETZ, T. A. *Plant/Operations Progress,* forthcoming.
16. *Control of Industrial Major Accident Hazard Regulations,* Statutory Instrument No. 1902, Her Majesty's Stationery Office, London, 1984.
17. European Community: Council Directive of 24 June 1982 on the Major Accident Hazards of Certain Industrial Activities, *Official Journal of the European Communities,* 1989, No. L230, 5 August 1982, pp. 1–18.
18. OFFORD, D. V. in *Her Majesty's Inspectors of Factories: Essays to Commemorate 150 Years of Health and Safety Inspection,* Her Majesty's Stationery Office, London, 1983, p. 57. This book gives a good account of the way in which the UK Factory Inspectorate, part of the Health and Safety Executive, goes about its work.

Chapter 9

Seveso

At Seveso near Milan in Italy, on 10 July 1976, a discharge of highly toxic dioxin from a bursting disc contaminated a neighbouring village. About 250 people developed the skin disease chloracne, and a large area of land was contaminated and made uninhabitable. Although no-one was killed it has become one of the best-known of all chemical plant accidents. It led to the enactment by the European Community of the Seveso Directive[1], which requires all companies which handle more than defined quantities of hazardous materials to demonstrate that they are capable of handling them safely. In the UK the Seveso Directive was implemented in the CIMAH (Control of Industrial Major Accident Hazard) Regulations[2]. No accident has resulted in greater 'legislative fall-out' than Seveso, although in the UK the CIMAH Regulations owe their origin to Flixborough and would have been enacted, in a slightly different form, if Seveso had never occurred.

The accident occurred on a batch plant for the manufacture of 2,4,5-trichlorophenol (TCP) from 1,2,4,5-tetrachlorobenzene and caustic soda, in the presence of ethylene glycol. TCP is used to make the herbicide 245T (2,4,5-trichlorophenoxyacetic acid). Dioxin (actually 2,3,7,8-tetrachlorodibenzodioxin) is not normally formed, except in minute amounts, but the reactor got too hot, a runaway reaction occurred, dioxin was formed and a rise in pressure caused the relief device to operate. There was no catchpot to collect the discharge, and the contents of the reactor, about 6 tonnes, including about 1 kg of dioxin, were distributed over the surrounding area.

How did the reactor get too hot?

Italian Law required the plant to shut down for the weekend, even though it was in the middle of a batch. On the weekend of the accident the operator had to shut it down at a stage, the end of the reaction but before removal of all the ethylene glycol, at which it had not been shut down before. Perhaps he had got behind with his duties. However, he had no reason to believe that it would be hazardous to shut it down at this stage. The reaction mixture was at 158°C, well below the temperature at which an exothermic reaction can start, believed at the time to be 230°C but possibly as low as 180°C. It was at first difficult to see how the reactor got hot enough for a runaway to start.

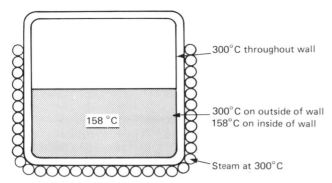

Figure 9.1   The Seveso reactor. The hot upper portion of the reactor wall heated the surface of the liquid

The reactor was heated by an external steam coil (Figure 9.1) which used exhaust steam, at a pressure of 12 bar and a temperature of about 190°C, from a turbine on another unit. The turbine was on reduced load, as units were shutting down for the weekend, and the temperature of the exhaust steam rose to about 300°C[3].

The temperature of the liquid could not get above its boiling point at the operating pressure, about 160°C, and so below the liquid level there was a temperature gradient through the reactor wall, the outside being at 300°C and the inside at 160°C. Above the liquid level the wall was at 300°C right through. When the steam was isolated, the temperature of the wall below the liquid level soon fell to that of the liquid, but above the liquid level it remained hotter. Heat passed from the upper wall to the surface of the liquid. When, 15 minutes later, the stirrer was switched off, the upper few centimetres of the liquid rose in temperature from 158°C to 180–190°C and a slow exothermic reaction started; after about seven hours a runaway occurred[4–6]. It is possible that the rise in temperature was catalysed by material which had been stuck to the upper wall of the reactor, was degraded by the heat and fell into the liquid.

The runaway would not have occurred if:

- Well-meaning legislators had not passed laws which left the management of the plant with no freedom to complete a batch before the weekend.
- The batch was not stopped at an unusual stage.
- A hazard and operability study (hazop) had been carried out on the design. Provided the service lines were studied as well as the process lines, under the heading 'more of temperature', this would probably have disclosed that when the turbine was on low load the temperature of the steam supplied to the reactor would rise. It is a well-established principle that whenever possible a heating medium should not be so hot that it can dangerously overheat the material being heated, and it was said to be a safety feature of the process used that there was a safety margin of 40°C between the temperature of the steam (190°C) and the runaway temperature (believed to be 230°C).

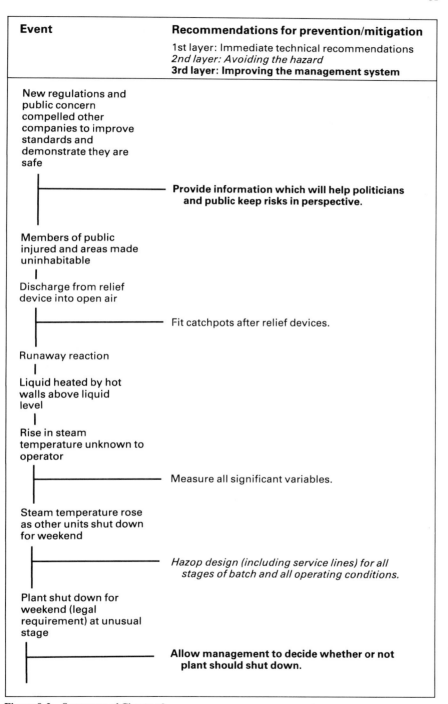

Event	Recommendations for prevention/mitigation
	1st layer: Immediate technical recommendations *2nd layer: Avoiding the hazard* **3rd layer: Improving the management system**
New regulations and public concern compelled other companies to improve standards and demonstrate they are safe	
	Provide information which will help politicians and public keep risks in perspective.
Members of public injured and areas made uninhabitable	
Discharge from relief device into open air	
	Fit catchpots after relief devices.
Runaway reaction	
Liquid heated by hot walls above liquid level	
Rise in steam temperature unknown to operator	
	Measure all significant variables.
Steam temperature rose as other units shut down for weekend	
	Hazop design (including service lines) for all stages of batch and all operating conditions.
Plant shut down for weekend (legal requirement) at unusual stage	
	Allow management to decide whether or not plant should shut down.

Figure 9.2 Summary of Chapter 9

It is, of course, necessary when carrying out a hazop on a batch plant to consider each deviation, such as more of temperature, for each stage of the batch. All hazops should include service lines as well as process lines.

During a hazop we ask if the operator will know if a deviation occurs. (This is important for operability as well as safety.) Asking the question would have drawn attention to the fact that the temperature of the steam was not measured, and a temperature measurement could have been added. In general, for every variable which can affect safety or operability significantly, we need a measurement. On many plants the temperature of the steam cannot vary significantly or it does not matter if it does. If it can vary significantly, and it does matter, it should be measured.

Why was there no relief system catchpot?

When hazardous materials may be discharged from relief devices or vents, it is normal practice to install a collection system; a flare system for flammable gases and vapours, a scrubbing system for toxic gases and vapours or a catchpot for solids and high-boiling liquids. Often more than one of these devices has to be provided. If a catchpot had been installed at Seveso the name would still be unknown.

It is possible that no catchpot was installed because the designers did not foresee that a runaway reaction could occur, and that the relief device was intended to protect against other causes of overpressure such as overfilling. However, it is not good practice to discharge even relatively harmless liquids into the open air and there had been three runaway reactions on other similar plants which should have been known to the designers. The worst occurred at Ludwigshaven in 1953; 55 men were affected by dioxin, some seriously, and it was found to be impossible to decontaminate the building, which had to be destroyed[6,7].

Public response

The public response was similar to that to Flixborough, rather less in the UK, rather greater in the rest of Europe, and it is not necessary to add anything to the discussion in Chapter 8.

References

1. European Community: Council Directive of 24 June 1982 on the Major Accident Hazards of Certain Industrial Activities, *Official Journal of the European Communities*, 1989, No. L230, 5 August 1982, pp. 1–18.
2. *Control of Industrial Major Accident Hazard Regulations*, Statutory Instrument No. 1902, Her Majesty's Stationery Office, London, 1984. See also *A Guide to the Control of Industrial Major Accident Hazard Regulations 1984*, Her Majesty's Stationery Office, London, 1985.
3. BRETHERICK, L. *Loss Prevention Bulletin*, No. 054, Dec. 1983, p. 38.
4. THEOFANOUS, T. G. *Nature*, Vol. 291, 25 June 1981, p. 640.
5. CARDILLO, P. et al., *Journal of Hazardous Materials*, Vol. 9, 1984, p. 221.
6. BRETHERICK, L. *Handbook of Reactive Chemical Hazards*, 3rd edition, Butterworths, 1985, p. 541.
7. LEES, F. P. *Loss Prevention in the Process Industries*, Butterworths, 1980, Vol. 2, Appendix 2.

Chapter 10

Bhopal

The worst accident in the history of the chemical industry occurred in the Union Carbide plant at Bhopal, India on 3 December 1984, when a leak of over 25 tonnes of methyl isocyanate (MIC) from a storage tank spread beyond the plant boundary, killing about 2000 people. The official figure was 1754 but according to some reports the true figure was nearer 10 000[1,2]. In addition about 200 000 people were injured. Most of those killed and injured were living in a shanty town that had grown up close to the plant.

Before 1984 the worst accidents that had occurred in the chemical industry were the explosion of a 50/50 mixture of ammonium sulphate and ammonium nitrate at Oppau in Germany in 1921, which killed 430 people, including 50 members of the public, and the explosion of a cargo of ammonium nitrate in a ship in Texas City harbour in 1947, which killed 552 people, mostly members of the public[3]. If conventional explosives are classified as chemicals then we should include the explosion of an ammunition ship in Halifax, Novia Scotia in 1917, which killed about 1800 people. However, earlier in 1984, on 19 November, 550 people were killed when a fire occurred at a liquefied petroleum gas processing plant and distribution centre in San Juanico, a suburb of Mexico City[4], and on 25 February at least 508 people, most of them children, were killed when a petrol pipe ruptured in Cubatao, near Sao Paulo, Brazil, and the petrol spread across a swamp and caught fire[5]. In both cases most of those killed were living in shanty towns. 1984 was thus a bad year for the chemical industry.

The Bhopal tragedy started when a tank of MIC – an intermediate used in the manufacture of the insecticide carbaryl – became contaminated with water, and a runaway reaction occurred. The temperature and pressure rose, the relief valve lifted and MIC vapour was discharged into the atmosphere. The protective equipment which should have prevented or minimized the discharge was out of action or not in full working order: the refrigeration system which should have cooled the storage tank was shut down, the scrubbing system which should have absorbed the vapour was not immediately available and the flare system, which should have burnt any vapour which got past the scrubbing system, was out of use.

'What you don't have, can't leak'

There are many lessons to be learnt from Bhopal but the most important is that the material which leaked need not have been there at all. It was an intermediate, not a product or raw material, and while it was convenient to store it, it was not essential to do so. Originally MIC was imported and had to be stored but later it was manufactured on site. Nevertheless over 100 tonnes were in store, some of it in drums. After the tragedy it was reported that Du Pont intended to eliminate intermediate storage from a similar plant that they operated. Instead they will use the MIC as soon as it is produced, so that instead of 40 tonnes in a tank there will be only 5–10 kg in a pipeline[6]. Mitsubishi were said to do this already[6, 7]. During the next few years other companies described plans for reducing their stocks of other hazardous intermediates[8]. At the end of 1985 Union Carbide claimed that the inventories of 36 toxic chemicals had been reduced by 74%[9].

The main lesson of Bhopal is thus the same as that of Flixborough: 'What you don't have, can't leak'. Whenever possible, we should reduce or eliminate inventories of hazardous materials, in process or storage. It is unfortunate, to say the least, that more notice was not taken of the papers written after Flixborough which stressed the desirability of inherently safer designs, as they are called[10–12]. It seems that most companies felt so confident of their ability to keep hazardous materials under control that they did not look for ways of reducing inventories. Yet to keep lions under control we need expensive added-on protective equipment which may fail or may be neglected. It is better to keep lambs instead.

Intermediate storage is usually wanted by plant managers, as it makes operation easier: one section of the plant can continue on-line when another is shut down. Computer studies on equipment availability always show that intermediate storage is needed. However, they do not allow for the fact that if intermediate storage is available it will always be used, and maintenance will be carried out at leisure, but if it is not available people do everything possible to keep the plant on-line, carrying out maintenance as soon as possible. The need for intermediate storage is a self-fulfilling prophecy.

Another alternative to intermediate storage is to build a slightly larger plant and accept fewer running hours per year. Intermediate storage, including working capital, is expensive, as well as hazardous, but this alternative is rarely considered.

If reducing inventories, or intensification as it is called, is not practicable an alternative is substitution, i.e. using a safer material or route. At Bhopal the product (carbaryl) was made from alpha-naphthol, methylamine and phosgene. The first two were reacted together to make MIC, which was then reacted with phosgene. In an alternative process used by the Israeli company Makhteshim, alpha-naphthol and phosgene are reacted together to make a chloroformate ester which is then reacted with methylamine to make carbaryl. The same raw materials are used but MIC is not formed at all (Figure 10.1)[13].

Of course, phosgene is also a hazardous material and its inventory should be kept as low as possible, or avoided altogether. If carbaryl can only be made via phosgene, perhaps another insecticide should be manufactured instead.

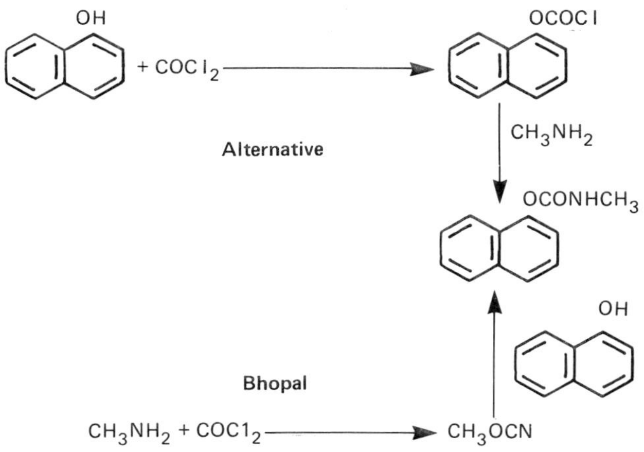

OH
+ COCl$_2$ ⟶ OCOCl

Alternative

| CH$_3$NH$_2$
↓

OCONHCH$_3$

OH

Bhopal

CH$_3$NH$_2$ + COCl$_2$ ⟶ CH$_3$OCN

Figure 10.1   Routes to carbaryl

Plant location

Just as materials which are not there cannot leak, people who are not there cannot be killed. The death toll at Bhopal would have been much smaller if a shanty town had not been allowed to grow up near the plant. In many countries planning controls prevent such developments, but not apparently in India. Of course, it is much more difficult to prevent the growth of shanty towns than of permanent dwellings, but nevertheless it is essential to stop them springing up close to hazardous plants. If the government of local authorities cannot control their growth then industry should be prepared, if necessary, to buy up the land and fence it off. As already mentioned, the high death tolls at Mexico City and Cubatao earlier in 1984 occurred in shanty towns.

It is generally agreed that if a leak of toxic gas occurs, people living in the path of the plume should stay indoors with their windows closed. Only if the leak is likely to continue for a long time, say more than half-an-hour, should evacuation be considered. However, this hardly applies to shanties, which are usually so well ventilated that the gas will enter them at once. It may be more difficult to prevent the growth of shanty towns than of permanent buildings, but it is also more important to do so.

Why did a runaway reaction occur?

The MIC storage tank became contaminated by substantial quantities of water and chloroform, up to a tonne of water and 1½ tonnes of chloroform, and this led to a complex series of runaway reactions, a rise in temperature and pressure and discharge of MIC vapour from the storage tank relief valves[14]. The precise route by which the water entered the tank is not known, though several theories have been put forward[15, 16]. One theory is that it came from a section of vent line some distance away that was being washed out. This vent line should have been isolated by a slip-plate (spade) but this had not been fitted. However, the water would have had to pass through six valves in series and it seems unlikely that a tonne could have entered the tank in this way. Another theory is that the water entered via the nitrogen supply line. Even sabotage has been suggested, but no supporting evidence has been supplied[17, 18]. The actual route hardly matters. If any of the suggested routes were possible, then they should have been closed before the disaster occurred. So far as is known no hazard and operability study (hazop)[19, 20] was carried out on the design though hazop is a powerful tool, used by many companies for many years, for identifying routes by which contamination (and other unwanted deviations) can occur.

Since it is well known that water reacts violently with MIC, no water should have been allowed anywhere near the equipment, for washing out lines or for any other purpose. If water is not there, it cannot leak in, no matter how many valves leak or how many errors are made. This is another example of inherently safer design.

Keep protective equipment in working order – and size it correctly

The storage tank was fitted with a refrigeration system but this was not in use. The scrubbing system which should have absorbed the MIC discharged through the relief valve was not in full working order. The flare system which should have flared any MIC which got past the scrubbing system was disconnected from the plant for repair. The high temperature and pressure on the MIC tank were at first ignored, as the instruments were poorly maintained and known to be unreliable[21]. The high-temperature alarm did not operate, as the set-point had been altered and was too high[22]. One of the main lessons from Bhopal is thus the need to keep all protective equipment in full working order.

It is easy to buy safety equipment; all we need is money and if we make enough fuss we get it in the end. It is much more difficult to make sure that the equipment is kept in full working order, especially when the initial enthusiasm has worn off. Procedures, including testing and maintenance procedures, are subject to a form of corrosion more rapid than that which affects the steelwork and can vanish without trace in a few months once managers lose interest. A continuous auditing effort is needed by managers at all levels to make sure that procedures are maintained. (See Chapter 6.)

Sometimes procedures lapse because managers lose interest. Unknown to them, operators discontinue safety measures. At Bhopal it went further than this. Disconnecting the flare system and shutting down the refrigeration system are hardly decisions that operators are likely to take on their own. The managers themselves must have taken these decisions and thus shown a lack of understanding and/or commitment.

It is possible that the protective equipment was out of use because the plant that produced the MIC was shut down and everyone assumed that the equipment had been installed to protect the plant rather than the storage. Runaway reactions, leaks, and discharges from relief valves are commoner on plants than on storage systems but they are by no means unknown on storage systems. Twenty-four of the 100 largest insurance losses in a 30 year period occurred in storage areas and their value was higher than average[23]. Furthermore, since a relief valve was installed on the storage tank, it was liable to lift and the protective equipment should have been available to handle the discharge. If the designers were sure that the relief valve would never lift there would have been no need to install it.

It has been argued that the refrigeration, scrubbing and flare systems were not designed to cope with a runaway reaction of the size that occurred and that there would have been a substantial discharge of MIC to atmosphere even if they had all been in full working order. This may be so, but they would certainly have reduced the size of the discharge and delayed its start.

The relief valve was too small for the discharge from a runaway reaction. The pressure in the storage vessel, designed for a gauge pressure of $40 \, lbf/in^2$ (2.7 bar), reached $200–250 \, lbf/in^2$ (14–17 bar). The vessel was distorted and nearly burst. If it had burst, the loss of life might have been lower, as there would have been less dispersion of the vapour. The relief valve was designed to handle vapour only but the actual flow was a two-phase mixture of vapour and liquid[24].

If the protective equipment was not designed to handle a runaway, or two-phase flow, we are entitled to ask why. Were the possibilities of a runaway or two-phase flow not foreseen or were they considered so unlikely that it was not necessary to guard against them? What formal procedures were used during design to answer these questions?

Although the managers (and also the operators, but they take their cue from the managers) showed less competence and commitment to safety than might reasonably have been expected, there is no reason to suppose that Indian managers (and operators) are in general less competent than those in the West. There are poor managers in every country. In one respect the managing director of Union Carbide India showed more awareness than his US colleagues: he queried the need for so much MIC storage but was overruled[15].

Bhopal illustrates the limitations of hazard assessment techniques. If asked before the accident to estimate the probability of a leak of MIC, by fault tree or similar techniques, most analysts would have estimated the failure rates of the refrigeration, scrubbing and flare systems but would not have considered the possibility that they might all be switched off. Hazard assessments become useless if the assumptions on which they are based are no longer true.

Similarly, estimates of human error are usually estimates of the probability that a person will forget to carry out a task, such as closing a valve, or carry it out wrongly. We cannot estimate the probability that he will make a conscious decision not to close it[25].

Some reports on Bhopal suggested that the instrumentation there was less sophisticated than on similar plants in the US and this may have led to the accident. This is a red herring. If conventional instrumentation was not adequately maintained and its warnings were ignored, then there is no reason to believe that computerized instrumentation would have been treated any differently. In fact the reverse may be the case. If people are unable or unwilling to maintain basic equipment, they are less likely to maintain sophisticated equipment. Nevertheless, during the investigation of accidents which have occurred because the safety equipment provided was not used, people often suggest that more equipment is installed. See Chapter 6.

Another protective device was a water spray system which was designed to absorb small leaks at or near ground level. It was not intended to absorb relief valve discharges at a high level and failed to do so.

Joint ventures

The Bhopal plant was half-owned by a US company, Union Carbide, and half-owned locally. Although Union Carbide had designed the plant and had been involved in the start-up, by the time of the accident the Indian company had become responsible for operations, as required by Indian law.

In the case of such joint ventures it is important to be clear as to who is responsible for safety in design and operation. The technically more sophisticated partner has a special responsibility, even if it is not directly responsible. It should make sure that the operating partner has the knowledge, skill, commitment and resources necessary for safe operation. If not, it should not go ahead with the venture. It cannot shrug off responsibility by saying that it is no longer in full control. Soon after Bhopal one commentator wrote, '. . . multinational companies and their host countries have got themselves into a situation in which neither feels fully responsible'[26]. If we give dangerous plant or material to people who have not demonstrated their competence to handle it we are responsible for the injuries they cause.

For Union Carbide the Bhopal plant was a backwater, making little contribution to profits, in fact often losing money, and may have received less than its fair share of management resources[27].

At Flixborough (Chapter 8) the partner with knowledge of the technology (Dutch State Mines) was in control.

Training in loss prevention

Bhopal makes us ask if the people in charge of the plant, and those who designed it, received sufficient training in loss prevention, as students and from their employers. In the UK all chemical engineering undergraduates receive some training in loss prevention[28]. If they do not they are not able to join the Institution of Chemical Engineers as full or corporate members. In most other countries, including the US, most undergraduate chemical engineers receive no such training, although the American Institute of Chemical Engineers is now encouraging universities to introduce it.

There are several reasons why loss prevention should be included in the training of chemical engineers[29].:

(1) Loss prevention should not be something added on to a plant after design, like a coat of paint, but an integral part of design. Hazards should, whenever possible, be removed by a change in design, such as reduction in inventory, rather than by adding on protective equipment. The designer should not ask the safety adviser to add on the safety

features for him; he should be taught to design a plant which does not require added-on safety features.

(2) Most engineers never use much of the knowledge they acquire as students but almost all have at some time to take decisions on loss prevention. Universities which give no training in loss prevention are not preparing their students for the tasks they will have to undertake.

(3) Loss prevention can be used to illustrate many of the principles of chemical engineering and to show that many problems which at first sight do not seem to lend themselves to numerical treatment can in fact be treated quantitatively.

Since, in many countries, universities are not providing training in chemical engineering, companies should make up the deficiency by internal training. Many try to but often in rather a haphazard way – an occasional course or lecture. Few companies put all new recruits through a planned programme.

At Bhopal the original managers had left and had been replaced by others whose experience had been mainly in the manufacture of batteries. There had been eight different managers in charge of the plant in 15 years[30]. Many of the original operators had also left and one wonders how well their successors were trained[31].

However, while these facts, and reductions in manning, may be evidence of poor management and a lack of commitment to safety, I do not think that they contributed directly to the accident. The errors that were made, such as disconnection of safety equipment and resetting trips at too high a level, were basic ones that cannot be blamed on inexperience of the particular plant. No manager who knew and accepted the first principles of loss prevention would have allowed them to occur.

Handling emergencies

More than any other accident described in this book, Bhopal showed up deficiencies in the procedures for handling emergencies, both by the company and by the local authorities. It showed clearly the need for companies to collaborate with the emergency services in identifying incidents that might occur and their consequences, drawing up plans to cope with them and exercising these plans. In the UK this is now required by law[32]. This aspect is discussed in reference 33.

Public response

In the US, Bhopal produced a public reaction similar to that produced in the UK by Flixborough (Chapter 8) and in the rest of Europe by Seveso (Chapter 9). Many companies spent a great deal of money and effort making sure that a similar accident could not occur on their plants. So far, however, Bhopal seems to have produced less 'regulatory fall-out' than Flixborough or Seveso. The US chemical industry has tried to convince the authorities that it can put its own house in order. In particular the American Institute of Chemical Engineers has set up a Center for Chemical Process Safety, generously funded by the chemical industry, to provide advice on loss prevention. One of its objectives is to have loss

Event	Recommendations for prevention/mitigation
	1st layer: Immediate technical recommendations *2nd layer: Avoiding the hazard* **3rd layer: Improving the management system**

Public concern compelled other companies to improve standards

Provide information that will help public keep risks in perspective.

Emergency not handled well

Provide and practise emergency plans.

About 2000 people killed

Control building near major hazards.

Scrubber not in full working order
Flare stack out of use
Both may have been too small

Keep protective equipment in working order. Size for foreseeable conditions.

Discharge from relief valve

Refrigeration system out of use

Keep protective equipment in use even though plant is shut down.

Runaway reaction

Rise in temperature

Train operators not to ignore unusual readings.

Water entered MIC tank

Carry out hazops on new designs.
Do not allow water near MIC.

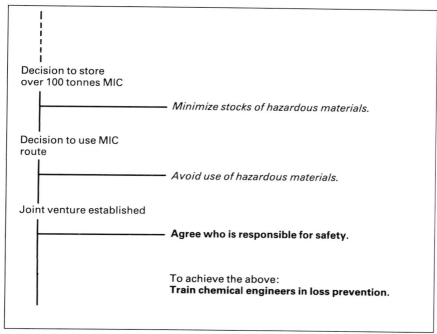

Figure 10.2   Summary of Chapter 10

prevention included in the training of undergraduates. The Chemical Manufacturers Association has launched a Community Awareness and Response (CAER) programme to encourage companies to improve their emergency plans and a National Chemical Response Information Center to provide the public and emergency services with advice and assistance before and during emergencies.

Nevertheless in a paper called 'A field day for the legislators', Stover[34] lists 32 US Government proposals or activities and 35 international activities that had been initiated by the end of 1985. In addition there have been state and local responses in the US.

In India there were, of course, extensive social effects. They are reviewed in reference 35.

Terrible though Bhopal was, we should beware of over-reaction and of suggestions that insecticides, or even the whole chemical industry, are unnecessary. Insecticides, by increasing food production, have saved far more people than Bhopal has killed. But Bhopal was not an inevitable result of insecticide manufacture. By better design or by better operation, by just one of the recommendations summarized in Figure 10.2, Bhopal could have been prevented. The most effective methods of prevention are those near the bottom of the diagram, such as reduction in inventory or change in the process. The safety measures at Bhopal, such as the scrubber and the flare stack, were too near the top of the chain, too near the top event. If they failed there was nothing to fall back on. To prevent the next Bhopal we need to start at the bottom of the chain.

References

1. *Free Labour World,* International Federation of Free Trade Unions, Brussels, Belgium, No. 1/86, 18 Jan. 1986, p. 1.
2. SHRIVASTAVA, P. *Bhopal – Anatomy of a Crisis,* Ballinger, Cambridge, Massachusetts, 1987, p. 64.
3. LEES, F. P. *Loss Prevention in the Process Industries,* Butterworths, 1980, Vol. 2, Appendix 3.
4. *Bleve – The Tragedy of San Juanico,* Skandia International, Stockholm, 1985.
5. *Hazardous Cargo Bulletin,* June 1984, p. 34.
6. *Chemical Week,* 23 Jan. 1985, p. 8.
7. *Chemistry in Britain,* Vol. 21, No. 2, Feb. 1985, p. 123.
8. WADE, D. E. *Proceedings of the International Symposium on Preventing Major Chemical Accidents, Washington, DC, 3–5 February 1987,* Paper 2.1.
9. *Chemical Insight,* Nov. 1985, p. 1.
10. KLETZ, T. A. *Chemical Engineering,* Vol. 83, No. 8, 12 April 1976, p. 124.
11. KLETZ, T. A. *Chemistry and Industry,* 6 May 1978, p. 37.
12. KLETZ, T. A. *Hydrocarbon Processing,* Vol. 59, No. 8, Aug. 1980, p. 137.
13. REUBEN, B. M. private communication.
14. *Bhopal Methyl Isocyanate Incident: Investigation Team Report,* Union Carbide Corporation, March 1985.
15. *The Trade Union Report on Bhopal,* International Federation of Free Trade Unions and International Federation of Chemical, Energy and General Workers' Unions, Geneva, Switzerland, 1985.
16. VARADARAJAN, S. *et al., Report on Scientific Studies on the Factors Related to Bhopal Toxic Gas Leakage,* Indian Planning Commission, Dec. 1985.
17. *New Scientist,* 14 Nov. 1985, p. 15.
18. As ref. 2, p. 50.
19. KLETZ, T. A. *Hazop and Hazan – Notes on the Identification and Assessment of Hazards,* Institution of Chemical Engineers, Rugby, UK, 2nd edition, 1986.
20. KLETZ, T. A. *Chemical Engineering,* Vol. 92, No. 7, 1 April 1985, p. 48.
21. *New York Times,* 28 Jan.–3 Feb. 1985.
22. SHRIVASTAVA, P. *The Accident at Union Carbide Plant in Bhopal – A Case Study,* Air Pollution Control Association Conference on Avoiding and Managing Environmental Damage from Major Industrial Accidents, Vancouver, Canada, 3–6 Nov. 1985.
23. *One Hundred Largest Losses,* Marsh and McLennan, Chicago, Illinois, 8th edition, 1985.
24. SWIFT, I. in *The Chemical Industry after Bhopal,* Proceedings of a Symposium, London, 7/8 Nov. 1985, IBC Technical Services.
25. KLETZ, T. A. *An Engineer's View of Human Error,* Institution of Chemical Engineers, Rugby, UK, 1985, Chapter 5.
26. SMITH, A. W. *The Daily Telegraph,* 15 Dec. 1984, p. 15.
27. As ref. 2, p. 51.
28. *A Scheme for a Degree Course in Chemical Engineering,* Institution of Chemical Engineers, Rugby, UK, 1983.
29. KLETZ, T. A. *Pipeline,* No. 25, Nov. 1986, p. 2, and *Plant/Operations Progress,* forthcoming.
30. As ref. 2, p. 52.
31. *The Bhopal Papers,* Transnationals Information Centre, London, 1986, p. 4.
32. *Control of Industrial Major Accident Hazard Regulations,* Statutory Instrument No. 1902, Her Majesty's Stationery Office, London, 1984.
33. As ref. 2, Chapter 6.
34. STOVER, W. in *The Chemical Industry after Bhopal,* Proceedings of a Symposium, London, 7/8 Nov. 1985, IBC Technical Services.
35. As ref. 2, Chapters 4 and 5.

Chapter 11

Three Mile Island

This chapter and the next one describe two nuclear accidents that have many lessons for the process industries. Recommendations of purely nuclear interest are not discussed.

On 28 March 1979 the nuclear power station at Three Mile Island in Pennsylvania overheated and a small amount of radioactivity escaped to the atmosphere. Even in the long term no-one is likely to be harmed but nevertheless the incident shattered public confidence in the nuclear industry and led to widespread demands for a halt to its expansion.

Figure 11.1 is a simplified diagram of a pressurized-water reactor (PWR), the type used at Three Mile Island and in most nuclear power

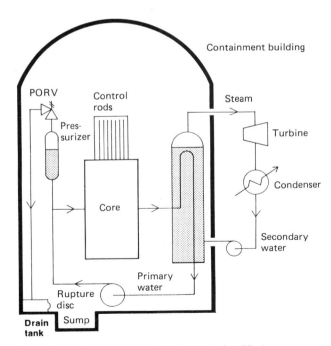

Figure 11.1 A pressurized-water reactor – simplified

stations. Existing UK reactors are gas-cooled but a PWR is being built at Sizewell. Heat is generated in the core by nuclear fission and is removed by pumping primary water round and round. The water is kept under pressure so that it is not boiling. (It is called a pressurized-water reactor to distinguish it from a boiling-water reactor, the type used at Chernobyl.) The primary water gives up its heat to the secondary water, which does boil. The steam produced drives a turbine and is condensed. The condensate is recycled. All radioactive materials, including the primary water, are enclosed in a containment building so that they will not escape if there is a leak[1-9].

Phase A – the trouble starts

The secondary water passed through a resin polisher unit to remove traces of impurities. There were several parallel paths and one of them choked. Less attention had been paid to the design of this off-the-shelf ancillary unit than to the design of the main radioactive equipment. In particular its reliability had not been studied to the same extent.

The first lesson to be learned, therefore, is that *packaged units, ancillary units, off-plots and so on need as much attention as the main stream,* especially when their failure can cause a shutdown of the main stream.

To try to clear the choke the operators used instrument air. Its pressure was lower than that of the water, so some water got back into the instrument air lines.

The second lesson is that we should *never connect service lines to process equipment at a higher pressure.* This has often been done and another case is described in Chapter 2, which also lists the precautions that should be taken if the pressures of the process or service lines are liable to change. In addition *it is bad practice to use instrument air lines for line blowing.* Ordinary compressed air should be used instead, as the results of any contamination of the air are then less serious.

There was a non-return valve in the compressed air line but it was faulty. Non-return valves have a poor reputation but in many plants they are never inspected or maintained. We cannot expect any piece of equipment, especially one containing moving parts, to operate correctly forever without attention. *Non-return valves should be scheduled for regular inspection,* say every year or two. However, non-return valves are designed to prevent gross flow, and should not be relied on to prevent a slight flow.

The water in the instrument air lines caused several instruments to fail and the turbine tripped. This stopped heat removal from the radioactive core. The production of heat by fission stopped automatically within a few minutes. (Metal rods dropped down into the core. They absorb neutrons and stop radioactive fission.) However, some heat was still produced by radioactive decay, about 6% of the total load, and this caused the primary water to boil. The pilot-operated relief valve (PORV) on the primary circuit lifted and pumps started up automatically to replace the water evaporated from the primary circuit.

Unfortunately the PORV stuck open.

Phase B – things worsen

The operators did not realize that the PORV was stuck open, because a light on the panel told them it was shut. However, the light was not operated by the valve position but by the signal to the valve. The operators had not been told this or had forgotten.

Whenever possible *instruments should measure directly what we want to know,* not some other property from which it can be inferred. If a direct measurement is impossible, then the label on the panel should tell us what is measured, in this case, 'Signal to PORV', not 'PORV position'.

Several other readings should have suggested to the operators that the PORV was stuck open and that the water in the primary circuit was boiling:

(1) The PORV exit line was hotter than usual, 140°C instead of 90°C, but this was thought to be due to residual heat.
(2) The pressure and temperature of the primary water were lower than usual.
(3) There was a high level in the containment building sump.
(4) The primary water circulation pumps were vibrating.

On the other hand the level in the pressurizer was high, as it was raised by bubbles of steam.

The operators decided to believe the PORV position light and the pressurizer level and ignore or explain away the other readings, probably because (a) they did not really understand how the temperature and pressure in the primary circuit depended on each other and when boiling would occur, and (b) their instructions and training had emphasized that it was dangerous to allow the primary circuit to get too full of water. They had not been told what to do if there was a small leak of primary water (although they had been told what to do if there was a major leak such as a pipe fracture).

The operators thought the PORV was shut. Conditions were clearly wrong and their training had emphasized that too much water was dangerous. They therefore shut down the water pumps. Note that the only action taken by the operators made matters worse. If they had done nothing, the plant would have cooled down safely on its own.

People sometimes say that automatic response is necessary if action has to be taken quickly but that we can rely on operators if they have plenty of time, say half-an-hour or more. However, this is true only if the operators can diagnose what is happening and know what action to take.

Three Mile Island shows us that we cannot operate complex plant by writing a series of precise instructions which must be followed to the letter. Problems will arise which are not foreseen or instruments will give conflicting results. Operators therefore need:

• To understand what goes on in the plant.
• To be trained in diagnostic skills.
• To be given diagnostic aids.

One method of diagnostic training has been described by Duncan and co-workers[10, 11]. The operator is shown a copy of the control panel on which certain readings are marked. He is asked to diagnose the fault and say what action he would take. The problems gradually increase in difficulty. The operator learns to identify and correct foreseeable faults and also learns general diagnostic skills that enable him to identify faults which have not been foreseen. Such training could have prevented Three Mile Island. (The fault which occurred there had not been foreseen though it should have been.)

Expert systems can be used to aid diagnosis[12].

Three Mile Island was not the first plant at which a PORV had stuck open. Similar incidents had happened before, though with less serious consequences. But unfortunately the lessons of these incidents had not been passed on. We pay a high price for the information that accidents give us – deaths, injuries and damage to plant. We should make full use of that knowledge.

Phase C – damage

With the make-up water isolated, the level in the primary circuit fell. The top of the radioactive core was uncovered, steam reacted with the zirconium alloy cans which protect the uranium, and hydrogen was formed. At the same time the steam which was being discharged through the PORV condensed in a drain tank, overflowed into the containment building sump and was automatically pumped outside the containment building.

Changes in design to minimize these consequences were recommended but they are not of general interest.

Damage did not occur until two hours after the start of the incident. The operators could have prevented the damage if, at any time during this period, they had realized that the PORV was open and that the water level was low. However, they had made their diagnosis and they stuck to it, even though the evidence against it was overwhelming. They had developed a mind-set.

This happens to us all. We have a problem. We think of an explanation, sometimes the first explanation that comes to mind. We are then so pleased with ourselves for solving the problem that we fail to see that our explanation is not fully satisfactory. Reference 13 describes some other accidents that occurred as the result of mind-sets. They are very difficult to avoid. Perhaps it would help if our training included examples of mind-sets so that we are aware that they occur and are thus a little less ready to grab the first explanation that we think of.

Following the rules in not enough

As already stated, it was believed at Three Mile Island, to a large extent, that all the operators had to do was to follow the rules. There was something of the same attitude amongst the managers and designers. They seem to have believed that if they followed the rules laid down by the US Nuclear Regulatory Commission they would have a safe plant. There is much less of this attitude in UK industry, where instead of a lot of detailed rules and regulations there is a general requirement, under the Health and Safety at Work Act, to provide a safe plant and system of work and adequate instruction, training and supervision, so far as is reasonably practicable. It is the responsibility of those who manage the plant, and know most about it, to say what is a safe plant etc., but if he does not agree the factory inspector will say so and, if necessary, issue an improvement or prohibition notice. If there is a generally accepted code of practice this should be followed, unless the employer can show that it is inapplicable or he is doing something else which is as safe or safer. Nevertheless, arguments in favour of more regulation surface from time to time in the UK so it may be worthwhile summarizing the case for the UK approach:

- Codes are more flexible than regulations. They can be changed more easily when new problems arise or new solutions are found to old problems.
- We do not have to follow to the letter regulations which are inappropriate or out-of-date.
- Managers cannot hide behind the regulations. They cannot say, 'My plant must be safe because I have followed all the rules'.
- It is not possible to write regulations to cover the details of complex and rapidly changing technologies.
- The factory inspector has more, not less, power under the UK system. He does not have to prove that a regulation has been broken. It is sufficient for him to say that the plant or method of working is not safe. His case is stronger if a generally accepted code of practice is not being followed.

In brief, *designers should have to demonstrate that their designs are safe, not just follow the rules.*

Consider minor failures as well as major ones

At Three Mile Island there was much concern with major failures, such as complete fracture of a primary water pipe, but smaller and more probable accidents were ignored. There was a belief that if large-scale accidents can be controlled, minor ones can be controlled as well. This is not true, and as a result what started as a minor incident became a major one. Similarly, in the process industries most injuries and damages are caused by minor failures, not major ones, and failures of both sorts should be considered. (See Chapter 15.)

Public response

Nobody was killed as a direct result of the accident at Three Mile Island but nevertheless it produced a greater public reaction than any other incident described in this book, with the exceptions of Bhopal and Chernobyl. The US nuclear industry received a setback from which it has not yet recovered, and in other countries it has been extensively quoted to show that the nuclear industry is not as competent as it had led us to believe and that improbable accidents can happen, so that it is better to stick to the hazards we know, such as coal mining. One of the lessons of Three Mile Island is that emergency plans should include plans for briefing the press and providing simple explanations of what has occurred and the extent of the risk.

Although no one was killed as a direct result of Three Mile Island nor by the subsequent radioactive fall-out, because the plant was shut down more coal had to be mined and burnt and this will have caused a few extra deaths by accident and pollution, perhaps two per year.

Must nuclear stations be so hazardous?

I have left to the end the most important lesson to be learned from Three Mile Island, one ignored by most commentators. The lesson is the same as that taught us by Bhopal (Chapter 10): whenever possible we should design plants that are inherently safe rather than plants which are made safe by adding on protective equipment which may fail or may be neglected. At Bhopal, instead of storing the intermediate MIC, it should have been used as soon as it was produced, or the product should have been manufactured by a route that did not involve MIC. In designing nuclear power stations we should prefer those designs which are least dependent on added-on engineered safety systems to prevent overheating of the core and damage to it if coolant flow is lost. The damage at Three Mile Island was, of course, much less serious than that which might have occurred or that which occurred at Chernobyl (Chapter 12).

Gas-cooled reactors are inherently safer than water-cooled ones because, if the coolant pressure is lost, convection cooling, assisted by a large mass of graphite, prevents serious overheating. In addition the operators have more time in which to act and are therefore less likely to make mistakes. Franklin writes, 'when operators are subject to conditions of extreme urgency . . . they will react in ways that lead to a high risk of promoting accidents rather than diminishing them. This is materially increased if operators are aware of the very small time margins that are available to them', and 'It is much better to have reactors which, even if they do not secure the last few percent of capital cost-effectiveness, provide the operator with half-an-hour to reflect on the consequences of the action before he needs to intervene'[14]. (At Three Mile Island the operators did have time, before damage occurred, to reconsider their decision to shut down the water pumps, but in general less time is available on water-cooled reactors than on gas-cooled ones.)

Other designs of reactors, still under development, are believed to be even more 'user-friendly' than gas-cooled ones. The sodium-cooled fast reactor can remove heat from the core by natural circulation if external power supplies fail. In the high-temperature gas reactor a small graphite-moderated core is cooled by high-pressure helium. The high temperature resistance of the fuel and the high surface-to-volume ratio ensure that the afterheat is lost by radiation and conduction. In the Swedish Process Inherent Ultimate Safety Reactor a water-cooled core is

Event	Recommendations for prevention/mitigation
	1st layer: Immediate technical recommendations *2nd layer: Avoiding the hazard* **3rd layer: Improving the management system**

Development of US
nuclear industry halted

Provide information that will help public keep risks in perspective.

Extreme public reaction

Plan how to handle public relations in emergencies.

Damage to core

Various design changes.

Water level fell
Operators shut down
water pumps

Pass on lessons of incidents elsewhere.

Operators believed
indicator light and
ignored other readings

**Train operators in diagnosis.
Provide diagnostic aids.
Train operators to understand, not just to follow the rules.
Consider minor failures as well as major ones.**

PORV stuck open
but indicator light
said it was shut

Measure directly what we need to know.

Primary water boiled
and PORV lifted

Turbine tripped

Instruments failed

Water entered
instrument air lines

Inspect NRVs regularly.

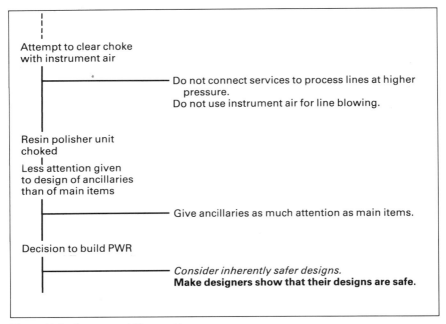

Figure 11.2   Summary of Chapter 11

immersed in boric acid solution. If the coolant pumps fail, the boric acid solution is drawn through the core by convection. The boron absorbs neutrons and stops the chain reaction while the water removes the residual heat. No make-up water is needed for a week[15].

In the short term the PWR, with its complex replicated added-on safety systems, may be the right answer for the UK, but the advantages of the inherently safer designs are so great that we may be building PWRs for only a few decades. For those countries that lack the resources, culture or commitment necessary to maintain complex added-on safety systems, PWRs are not the answer even today and they should wait until inherently safer designs are available.

References

This chapter is based on an article published in *Hydrocarbon Processing,* Vol. 61, No. 6, June 1982, p. 187, and thanks are due to Gulf Publishing Co. for their permission to quote from it. The text is based mainly on references 1–9.

1. BROOKS, G. L. and SIDDALL, E. *An Analysis of the Three Mile Island Accident,* CNS First Annual Conference, Montreal, Canada, 18 June 1980.
2. LEWIS, H. W. *Science,* March 1980, p. 33.
3. LANOUETTE, W. J. *The Bulletin of the Atomic Scientists,* Jan. 1980, p. 20.
4. SCHNEIDER, A. *Loss Prevention,* Vol. 14, 1981, p. 96.
5. *Report of the President's Commission on the Accident at Three Mile Island (The Kemeny Report),* Pergamon Press, 1979.
6. *TMI-2 Lessons Learned Task Force Report,* Report No. NUREG-0585, US Nuclear Regulatory Commission, 1979.

7. *Investigation into the March 28, 1979 Three Mile Island Accident by the Office of Inspection and Enforcement,* Report No. 50-320/79-10, US Nuclear Regulatory Commission, 1979.

8. *Accident at the Three Mile Island Nuclear Power Plant: Oversight hearings before a task force of the Subcommittee on Energy and the Environment of the Committee on Interior and Insular Affairs of the House of Representatives,* Serial No. 96-8, 1979.

9. MOSS, T. H. and SILLS, D. L. (editors), *The Three Mile Island Nuclear Accident: Lessons and Implications,* Annals of the New York Academy of Science, Vol. 365, 1981.

10. MARSHALL, E. E. *et al., The Chemical Engineer,* Feb. 1981, p. 66.

11. DUNCAN, K. D. *New Technology and Human Error,* edited by J. Rasmussen, K. D. Duncan and J. Leplat, Wiley, 1987, Chapter 19.

12. ANDOW, P. K. *Plant/Operations Progress,* Vol. 4, No. 2, April 1985, p. 116.

13. KLETZ, T. A. *An Engineer's View of Human Error,* Institution of Chemical Engineers, Rugby, UK, 1985, p. 56.

14. FRANKLIN, N. *The Chemical Engineer,* No. 430, Nov. 1986, p. 17.

15. WEINBERG, A. M. and SPIEWAK, I. *Science,* Vol. 224, No. 4656, 29 June 1984, p. 1398.

Chapter 12

Chernobyl

Like the accident at Three Mile Island (Chapter 11), the accident at Chernobyl, in the Ukraine, on 26 April, 1986, was the result of overheating of a water-cooled nuclear reactor, but there the similarity ends. Although both accidents were said to be due to human error, Three Mile Island was due to a failure to understand what was going on, the result of inadequate training, while Chernobyl was the result of a deliberate decision to ignore the normal safety instructions and override the normal protective equipment. At Three Mile Island the discharge of radioactive material is thought to have been too small to cause any harm. At Chernobyl about 30 people were killed immediately and it has been estimated that several thousand more will die from cancer during the next 30 years[1]. (For comparison, about 30 million people will die from cancer in Europe during this period.) However, this figure is subject to considerable uncertainty. It assumes that the effects of small doses of radiation are proportional to the dose, an assumption that may not be correct, and that no significant advances in the treatment of cancer will be made.

The reactor and the experiment

At Three Mile Island the water that cooled the reactor was kept under pressure and did not boil. In contrast, the Chernobyl reactor was cooled by boiling water. About half the Soviet reactors are of this type; the design is one that is not used outside the USSR, though other designs of boiling water reactor are used.

Figure 12.1 is a simplified diagram of the Chernobyl reactor. The uranium fuel is contained in 1661 pressure tubes, made of zirconium alloy, through which the water is passed. About 15% of the water is turned into steam, which is separated from the water in a steam drum and drives a turbine. The exhaust steam is condensed and recycled. Note that there is only one water stream – not two, as in a pressurized-water reactor. The pressure tubes are located in a large block of graphite which acts as a moderator, i.e. it slows the neutrons down so that they will react with the uranium. The graphite block is surrounded by a concrete shield, 2 metres thick at the side and 3 metres thick on top. About 200 control rods can move in and out of the reactor to absorb neutrons and thus control the rate at which heat is produced and to shut down the reactor in an emergency.

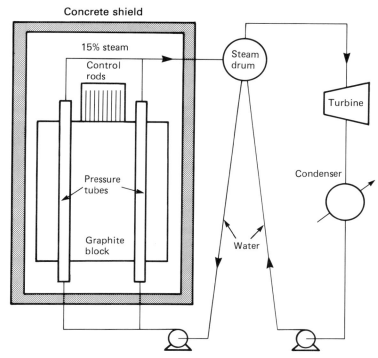

Figure 12.1   The Chernobyl reactor, simplified. The fuel is contained in 1661 zirconium alloy pressure tubes set in a graphite block. Water passes through the tubes. There are over 200 control rods

If reaction is stopped, heat is still produced, at about 6% of the normal rate, by radioactive decay, so cooling has to be kept in operation or the reactor will overheat (as in a pressurized-water reactor). As the reactor is not now producing power, the power to drive the water pumps and other auxiliary equipment has to come from the grid. Diesel generators are provided in case the grid supply is not available. They take about a minute to start up. During this time there may be no power available. The managers of the power station wanted to know if the turbine, as it slows down, will generate enough power to keep the auxiliaries on line. An experiment had shown that it would not but some changes had been made to the electrical equipment and they decided to repeat the experiment when the reactor was shutting down for a planned overhaul.

The experiment was a perfectly reasonable one to undertake and had been carried out before without incident, but now we come to the fatal error of judgement. The managers wanted to carry out the experiment several times. If they shut the reactor down it would take time to get it back on line, time that was not available, so they decided to keep the reactor on low rate and simply isolate the steam supply to the turbine. The reactor was steadied out, with some difficulty, at 6% of full output, although instructions said that it should not be operated at less than 20% output, and it was intended to carry out the experiment at 25% output. It took a long time to get it steadied out at this low rate – the test had to be postponed as power was required urgently – and during operation under these abnormal conditions there was a build-up of xenon in the core. Xenon is formed during operation at low rates and is used up at high rates. It is a 'poison', as it absorbs neutrons and slows down the rate of nuclear fission, and therefore most of the control rods had to be withdrawn to keep the reactor running. In addition, because of the low rate, the cooling system contained few steam bubbles and the water absorbed more neutrons. As a result the equivalent of only six rods were fully inserted though the instructions stated that there should always be at least 30. Finally the automatic trip system, which shuts down the reactor if there is a fault, was isolated. There were thus at least three deliberate decisions to depart from the safety instructions.

The experiment goes wrong

The experiment was not successful. The turbine, as it slowed down, did not produce enough power to keep the water pumps running at full speed and the reactor temperature started to rise. The automatic trip should have dropped the control rods and stopped the nuclear reaction but it had been switched off. The operators tried to insert the control rods by manual control but they could not insert them quickly enough, the reactor temperature continued to rise, the heat production increasing 100-fold in one second, and the temperature reached 3000–4000°C.

The temperature rose so rapidly because at low output rates, below 20%, the reactor had a positive power coefficient, i.e. as the temperature rose the rate of heat production also rose. (No other commercial reactor anywhere in the world has this feature.) For this reason instructions stated quite clearly that the reactor should never be operated below 20% load. In fact, it is doubtful if even the automatic equipment could have inserted the control rods quickly enough at low output rates. As the temperature rose, the water boiled more vigorously, more bubbles were formed and the heat transfer coefficient fell. The rising temperature in the fuel caused an increase in its internal pressure, as it contained volatile fission products, and this caused the fuel to disintegrate. The steam produced burst some of the pressure tubes, damaged the graphite and blew the 3 metre thick concrete shield off the top of the reactor. The building was damaged and fuel and fission products discharged into the atmosphere. About 30 people were killed immediately or died within a few weeks and, as already stated, many more will probably die in the years to come. Over 100 000 people were evacuated from an area 30 km in radius around the reactor, and part of this area will be unfit to live in for many years. The fission products were detected in most European countries.

The lessons to be learned

The first and most obvious lesson is that *protective equipment should not be isolated and basic safety rules should not be ignored.* In addition, as the consequences of breaking the rules were so serious, *the plant should have been designed so that the rules could not be ignored,* i.e. so that the automatic trip could not be isolated, so that at least 30 control rods had to be left in the core and so that the plant could not remain on-line if the output was reduced below 20%. The Soviet designers seem to have assumed that instructions would always be obeyed and that therefore there was no need for them to install protective equipment to prevent unsafe methods of operation. They assumed that the operators would be more reliable than automatic equipment, but they were not. (This is not a universal rule. Sometimes operators are more reliable than automatic equipment, but not in this case.)

Russians have a reputation for always following the rules and referring every detail to higher authority so it is at first sight surprising that they broke so many rules at Chernobyl. They seem to have had contradictory instructions – to carry out the tests as quickly and effectively as possible and to follow normal operating instructions – and they seem to have assumed that the instruction to carry out the experiment overrode the normal safety instructions. In the process industries managers have often had to argue with research workers who wanted to carry out experiments on the plant but did not want to be bound by the normal safety instructions. *Safety instructions should be followed at all times unless an exception has been authorized at the appropriate level after a systematic consideration of the hazards.* In addition, managers who talk a lot about output or efficiency or getting things done, without any mention of safety, inevitably leave operators with the impression that safety is less important. Managers should remember, when giving instructions, that *what you don't say is as important as what you do say.*

It is, of course, possible that operators had been breaking rules for some time – when everything has to be referred to the top it may be necessary to break the rules in order to get anything done – in which case regular audits would have picked it up. The USSR has no independent inspection service such as the UK Nuclear Installations Inspectorate.

None of the reports on Chernobyl make it clear at what level the decisions to operate at low rate, withdraw most of the control rods and switch off the trip were taken. It is clear, however, that when the normal rules were suspended, or the operators thought they were, they had insufficient understanding of the process to be able to replace rule-based behaviour by skill-based behaviour. At Three Mile Island, when an unforeseen fault occurred, the operators were similarly unable to cope.

Rule-based behaviour is appropriate when straightforward tasks have to be performed, but process plants do not come into this category. (In theory, if rule-based behaviour is all that is needed a computer can be used instead of a person.) On process plants we should never rely solely on rule-based behaviour, as circumstances may arise which were not foreseen by those who wrote the rules. Note also that skill-based behaviour requires motivation as well as skill. The operators may not have understood why it was so important to operate above 20% output. Like many operators, they may have relied more on process feel than theoretical knowledge and felt confident of their ability to control the reactor under all circumstances.

On process plants protective equipment sometimes has to be isolated, e.g. if it goes out of order or if the plant is operating under abnormal conditions – a low flow trip may have to be isolated during start-up – but they should be isolated only after authorization at an appropriate level, and the fact that the trip is isolated should be signalled in some way, e.g. by a light on the panel. If possible the trip should reset itself after a period of time. When the potential consequences of a hazard are serious, as in a nuclear power station, isolation of a trip should be impossible. High reliability is then essential and can be achieved by redundancy. Similarly, it is neither necessary nor possible to install protective equipment to prevent operators breaking every safety rule, but certain modes of operation may have such serious consequences that it should be impossible to adopt them.

The managers do not seem to have asked themselves what would occur if the experiment went wrong. *Before every experiment we should list all possible outcomes and their effects, and decide how they will be handled.*

Underlying the above was a basic ignorance of the first principles of loss prevention amongst designers and managers, and the comments made in Chapter 10 on Bhopal apply even more strongly to Chernobyl. The Russians now agree that there were errors in the design of the reactor and they have made changes to their other reactors, e.g. they have limited the extent to which control rods can be withdrawn and these are now kept in a region of high neutron flux so that their movement will have an immediate effect. In addition, they have made it harder for the protective systems to be defeated. They also agree that there were errors of judgement by the managers, who have been prosecuted and sent to prison, and that the training of the operators was inadequate, but it is not clear that they recognize the need for changes in their approach to loss prevention.

The need for inherently safer designs

The comments made in Chapter 11 on Three Mile Island apply with greater force to Chernobyl. In addition to the need for emergency cooling systems, which is common to all water-cooled reactors, the Chernobyl design was inherently less safe than the reactor at Three Mile Island. As already stated, if it got hotter the rate of heat production increased. In addition, at temperatures above 1000°C the water reacted with the zirconium pressure tubes, generating hydrogen. The hydrogen may have caught fire after the top was blown off the reactor. Once the pressure tubes burst, the water may have reacted with the graphite, also producing hydrogen, but this is an endothermic reaction and therefore less serious than the zirconium–water reaction. Whenever possible we should *avoid using in close proximity materials which react with each other,* even though reaction does not occur until the temperature rises, a container leaks or there is some other departure from normal operating conditions.

Another weakness of the Chernobyl reactor was its susceptibility to knock-on effects. Once a number of pressure tubes burst – considered an 'incredible' event by the designers – other tubes and the graphite were damaged and the concrete top was blown off the reactor. The roof of the building was flammable and caught fire.

Event	Recommendations for prevention/mitigation
	1st layer: Immediate technical recommendations *2nd layer: Avoiding the hazard* **3rd layer: Improving the management system**

Loss of confidence
in nuclear industry

————————————— **Explain differences in design between Chernobyl and other reactors.**

30 people killed,
many more at risk,
100 000 evacuated

Several tonnes of fuel
and fission products
escape

Top blown off reactor

————————————— *Design reactors less prone to knock-on effects.*

Pressure tubes burst

Water and Zr react

————————————— *Do not use materials which react with each other.*

Overheating gets worse

————————————— Increase number of control rods and speed of insertion.
Do not build reactors with positive power coefficients.

Control rods inserted
on manual control

Reactor overheats

————————————— *Before experiments, ask what can happen and make plans.*

Reactor did not trip

————————————— *Design trips that cannot be isolated.*

As turbine slowed,
pumps slowed

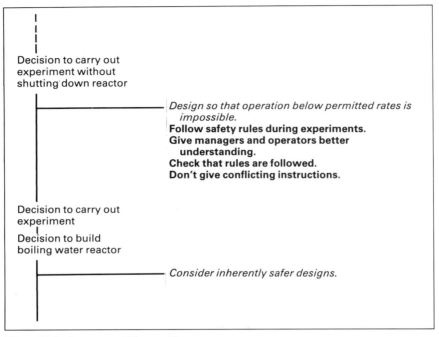

Decision to carry out
experiment without
shutting down reactor

Design so that operation below permitted rates is impossible.
Follow safety rules during experiments.
Give managers and operators better understanding.
Check that rules are followed.
Don't give conflicting instructions.

Decision to carry out
experiment

Decision to build
boiling water reactor

Consider inherently safer designs.

Figure 12.2   Summary of Chapter 12

Public response

Chernobyl was by far the worst nuclear accident that has occurred, though comparable with Bhopal if the probable long-term results are taken into account, and the public response was correspondingly great. The official attitude in the West was that the design of the reactor, and licensing procedures, were so different from those in use elsewhere that there was no reason to doubt the safety of reactors in other countries, but the public and press remained sceptical. Though the public is inclined to treat the nuclear industry as one, it does not always treat other industries in the same way. Collapse of a dam in India does not make people doubt the safety of UK dams or result in demands for them to be phased out. Perhaps the difference is due to the fact that the discharge from Chernobyl spread across national boundaries. (London is more exposed to fall-out from French and Belgian reactors than from UK ones.) While the detailed lessons of Chernobyl have been studied in the West, there does not seem to have been any major change in designs or procedures, legislative or otherwise. In this respect it has had less effect than Flixborough, Seveso or Bhopal. However, in the long run Chernobyl may encourage the development of inherently safer or more user-friendly reactors, as discussed in Chapter 11, and the need for the nuclear industry to sell itself to the public is now greater than ever.

However, we should kept the effects of Chernobyl in perspective. If a UK smoker who is worried about the effects smokes one cigarette less – not one per day or per year but one once and for all – he will have reduced his risk of getting cancer more than Chernobyl has increased it.

Arrangements in the UK for monitoring fall-out and its accumulation in food and for informing and advising the public were found to be lacking, and many changes have been made in this area.

References

The text is based mainly on the following:
1. *Atom,* No. 366, April 1977, p. 16.
2. GITTUS, J. and DALGLISH, J. *Atom,* No. 360, Oct. 1986, p. 6.
3. GITTUS, J. *Atom,* No. 386, June 1987, p. 2.
4. FRANKLIN, N. *The Chemical Engineer,* No. 430, Nov. 1986, p. 17.
5. *Physics Bulletin,* Vol. 37, No. 11, Nov. 1986, p. 447.
6. HOHENEMSER, C. *et al., Chem. Tech.,* Oct. 1986, p. 596.
7. ROWLAND, P. R. *Chemistry in Britain,* Vol. 22, No. 10, Oct. 1986, p. 907.
8. GITTUS, J. H. *et al., The Chernobyl Accident and its Consequences,* Her Majesty's Stationery Office, London, 1987.
9. REASON, J. *Bulletin of the British Psychological Society,* Vol. 40, 1987, p. 201.

Chapter 13

Aberfan

On 21 October 1966 a waste tip from a coal mine collapsed onto the village of Aberfan in South Wales. A school lay in the path of the waste and the 166 people killed were mainly children. Compared with the other accidents described in this book, the immediate technical causes of the disaster were simple – the tip was constructed above a stream on sloping ground – but the official report[1] brought out very clearly the underlying weakness in the management system, namely a failure to learn from the past, a failure to inspect regularly and a failure to employ competent and well-trained people with the right professional qualifications. The official report, on which this chapter is based, also includes many percipient remarks which apply to accidents generally. Quotations from the report are in *italics* and the references are to paragraph numbers.

A failure to learn from the past

. . . forty years before it occurred, we have the basic cause of the Aberfan disaster being recognised and warned against. But, as we shall see, it was a warning which went largely unheeded. (¶44)

In 1927 Professor George Knox presented a paper on 'Landslips in South Wales Valleys' which became widely known at the time. He gave full warning of the menace to tip stability presented by water and said that 'if you do not pay for drainage, you will have to pay for landslips in other ways'.

Tip slides are not new phenomena. Although not frequent, they have happened throughout the world and particularly in South Wales for many years, and they have given rise to quite an extensive body of literature available long before the disaster. (¶72) In 1939 there was a tip slide at Cilfynydd in South Wales. *Its speed and destructive effect were comparable with the disaster at Aberfan, but fortunately no school, house or person lay in its path. . . . It could not fail to have alerted the minds of all reasonably prudent personnel employed in the industry of the dangers lurking in coal-tips . . . the lesson if ever learnt, was soon forgotten.* (¶82)

In 1944 another tip at Aberfan slid 500–600 feet. Apparently, no-one troubled to investigate why it had slipped, but a completely adequate explanation was to hand . . . it covered 400 feet of a totally unculverted stream. (¶88) *To all who had eyes to see, it provided a constant and vivid reminder (if any were needed) that tips built on slopes can and do slip and, once having started, can and do travel long distances. (¶89)*

Why was there this general neglect? Human nature being what it is, we think the answer to this question lies in the fact that . . . there is no previous case of loss of life due to tip stability . . . (¶68). The Inspectorate of Mines and Quarries were not even informed of the 1944 slide, as no-one was killed or injured, and they never looked on tip stability as a problem meriting close inspection or recommendation.

Aberfan therefore shows very clearly that we should learn from all accidents, those that could have caused death or injury as well as those that did, and that a conscious management effort is needed to make sure that the lessons of the past are not forgotten.

A failure to inspect adequately

The simple truth is that there was no regular inspection of the tips. On the contrary, their inspection (such as it was) was wholly haphazard in point of time and had no reference to their stability at all, but simply related to such matters as the condition of the mechanical equipment for tipping. (¶64)

In 1965 there was a tip slide at Tymawr Colliery in South Wales. A main road was flooded and two or three cars in the colliery car park were damaged. The Chief Engineer of the South Western Division of the National Coal Board sent a memorandum, 'Precautions to Prevent Sliding', an update of one issued after the 1939 slide, to the Area Chief Engineers, Mechanical Engineers and Civil Engineers under his control. In the covering letter he said, *'I should be pleased, therefore, if you would arrange with your colleagues, for a detailed examination of every tip within your Area, and to take the necessary action for its immediate safety and ultimate good management'. (¶162)*

The inspection of the Aberfan tips was carried out by the Area Mechanical Engineer. . . . it was remarkably inadequate. It was purely visual; (he) had no plans or surveys with him and he made no notes . . . and he never climbed to the top of Tip 7 to see conditions for himself or to seek information from the tipping gang . . . There was in truth much for him to note – the extent of the 1944 slip, the results of the incident of 1963, and the deep bowl-like depression, to mention just some of the unusual features . . . Had these obvious features been noted and properly enquired into, it seems inconceivable that there would have been a disaster. (¶170)

A failure to employ competent and well-trained people

The Area Mechanical Engineer was strongly criticized by the Enquiry (*for failing to exercise anything like proper care in the manner in which he purported to discharge the duty of inspection laid upon him* [¶171]) but he was a mechanical engineer, not a civil, and had had no training to qualify him for the task of tip inspection. In addition, like the engineers who built the pipe which failed at Flixborough (Chapter 8), he had no professional training or qualification but had started as an apprentice at the age of 15. Like them he did not know what he did not know.

It was customary in the coal industry for tips to be the responsibility of mechanical rather than civil engineers. *It was left to the mechanical engineer to do with tips what seemed best to him in such time as was available after attending to his multifarious other duties.* (¶70) *For our part, we are firmly of the opinion that, had a competent civil engineer examined Tip 7 in May, 1965, the inevitable result would have been that all tipping there would have been stopped.* (¶73) *The tragedy is that no civil engineer ever examined it and that vast quantities of refuse continued to be tipped there for another 18 months.* (¶168)

Why did the Area Civil Engineer, who also received the 1965 memorandum, not take an interest in the tip? It was *traditional that tips were the responsibility of mechanical engineers.* In addition, the Civil Engineer was heavily overshadowed by the far more dominant Mechanical Engineer and took it for granted that his assistance was not wanted. (¶165)

Tip 7 was started at Easter 1958 . . . The two men who took the decision to start it were . . . *quite unfitted by training to come to an unaided decision as to the suitability of the proposed site . . . They had no Ordnance Survey map, and they took no plan with them, because none existed; they made no boreholes; they came to no conclusion regarding the limits of the tipping area; and they consulted no one else – not even the Colliery Surveyor. They arranged for no drainage, for they considered none necessary. It was a case of the blind leading the blind . . . in a system which had been inherited from the blind.* (¶104) *In our judgment, such inspection as they made was worthless. They were unfitted by training to judge the matter, and what stared them in the face they ignored.* (¶105)

Personal responsibility

We found that many witnesses, not excluding those who were intelligent and anxious to assist us, had been oblivious of what lay before their eyes. It did not enter their consciousness. They were like moles being asked about the habits of birds. (¶17)

. . . the Report which follows tells not of wickedness but of ignorance, ineptitude and a failure in communications. Ignorance on the part of those charged at all levels with the siting, control and daily management of tips; bungling ineptitude on the part of those who had the duty of supervising and directing them; and failure on the part of those having knowledge of the factors which affect tip stability to communicate that knowledge and to see that it was applied. (¶18)

The stark truth is that the tragedy of Aberfan flowed from the fact that, notwithstanding the lessons of the recent past, not for one fleeting moment did many otherwise conscientious and able men turn their attention to the problem of tip stability . . . These men were not thinking or working in a vacuum. All that was required of them was a sober and intelligent consideration of the established facts. (¶46)

. . . the Aberfan disaster is a terrifying tale of bungling ineptitude by many men charged with tasks for which they were totally unfitted, of failure to heed clear warnings, and of a total lack of direction from above. Not villains, but decent men, led astray by foolishness or by ignorance or by both in combination, are responsible for what happened at Aberfan. (¶47)

The incidents preceding the disaster . . . should, in our judgement, have served . . . to bring home vividly to all having any interest in coal-mining that tips placed on a hillside can and do slip and, having started, can move quickly and far; that it was accordingly necessary to formulate and maintain a system aimed at preventing such a happening; and for that purpose to issue directions, disseminate information, train personnel, inspect frequently and report regularly. These events were so spread out over the years that there was ample time for their significance to be reflected upon and realised and so to lead to effective action. But the bitter truth is that they were allowed to pass unheeded into the limbo of forgotten things. (¶48)

. . . most of the men whose acts and omissions we have had to consider have had, as it were, a bad upbringing. They have not been taught to be cautious, they were not made aware of any need for caution, they were left uninformed as to the tell-tale signs on a tip which should have alerted them. Accordingly, if in the last analysis, any of them must be blamed individually for contributing to the disaster . . . a strong 'plea in mitigation' may be advanced . . . personal responsibility is subordinate to confessed failure to have a policy governing tip stability . . . It is in the realm of an absence of policy that the gravest strictures lie, and it is that absence which must be the root cause of the disaster. (¶182)

Change a few words and the passages just quoted would apply to many more of the accidents discussed in this book, and many others.

The local Member of Parliament admitted at the Enquiry that he entertained the thought that the tip 'might not only slide but in sliding reach the village' but did nothing as he feared that the result of drawing attention to the tip might be closure of the colliery. (¶61) He presumably did not realize how serious the consequences of a slide might be but nevertheless it reminds us that anyone, especially someone in a position of authority, should speak up if he sees a hazard. If we see a hole in the road, then morally (though not legally, . . . *the law recognizes no* general *duty to protect another from harm* (¶771)) we are obliged to prevent other people falling in it. See page 70.

Inherently safer designs

Chapters 10–12 have stressed that whenever possible we should try to avoid hazards rather than control them. This is not discussed in the official Report but applying that philosophy leads to the suggestion that whenever possible we should locate tips so that, if they do slide, casualties are likely to be small.

Going further, what you don't have can't slide. The less coal we use, the fewer tips we shall need. Increased use of natural gas and nuclear electricity will decrease our need for coal. We can hardly blame the National Coal Board for not pointing this out, especially as most of the tips in existence were started before natural gas and nuclear electricity became available. However, the opponents of nuclear electricity make much of the hazards of nuclear waste but the quantities involved are tiny compared with the waste produced by coal mining. (If all electricity was made from nuclear energy, instead of 20% as in the UK, then each person in his lifetime would account for a piece of highly active waste the size of an

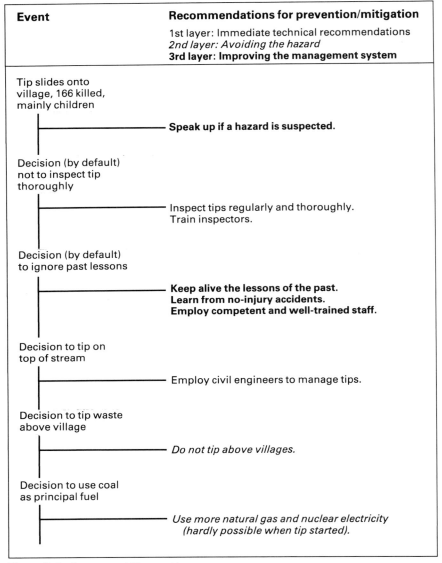

Figure 13.1   Summary of Chapter 13

orange.) They do not remind us of Aberfan but it is just as relevant to point out what actually happened at Aberfan as what might happen in the disposal of nuclear waste. Of course, coal-mining tips can be managed safely, but so can nuclear waste disposal.

Reference

1. *Report of the Tribunal appointed to inquire into the Disaster at Aberfan on October 21st, 1966*, Her Majesty's Stationery Office, London, 1967. Thanks are due to the Controller of Her Majesty's Stationery Office for permission to quote from this report.

Chapter 14

Missing recommendations

We should include in accident reports all the facts that have come to light, even though no conclusions are drawn from some of them, so that readers with different backgrounds, experience or interests can draw additional conclusions which were not obvious to the original investigators.

For example, the official report on Flixborough (Chapter 8), though outstanding in many ways, did not discuss inherently safer designs. But the report did contain the information from which others could draw an important additional conclusion: the most effective way of preventing similar accidents in the future is to design plants which do not contain such large inventories of hazardous materials.

This chapter summarizes three accident reports from which some readers were able to draw conclusions not apparent to the original authors. Two of the reports are UK official ones but this does not imply that the authors of official reports are poor at drawing conclusions. UK official reports include much information from which some readers can draw additional conclusions.

Perhaps the most extreme example of a report which failed to draw the necessary conclusions was one which said that there was no need to recommend any changes in design, as the plant had been destroyed!

A tank explosion

A hydrocarbon storage tank had to be gas-freed for entry. Air was blown into it with a fan. The vapour/air mixture in the tank had to pass through the explosive range. The fan impellor disintegrated and the resulting sparks ignited the vapour/air mixture. An explosion occurred.

The recommendation was to use a better fan[1].

Most readers will have drawn the conclusion that tanks should be gas-freed with nitrogen, steam or an inert gas. If air is used an explosive mixture will be formed, and experience shows that sooner or later a source of ignition will turn up. Using a better fan merely ensures that one particular source of ignition is less likely to turn up again. It does not deal with the underlying problem, that flammable mixtures are potentially hazardous and should never be permitted except under rigidly defined conditions where the risk of an explosion can be accepted.

The unforeseen results of technical change

Eleven men were killed in an explosion at a steelworks in Scunthorpe, Lincolnshire in 1975. Molten iron at 1500°C was being run out of a blast furnace into a closed ladle or torpedo, a long, thin vessel mounted on a railway truck and able to carry 175 tonnes of iron. The molten iron entered the torpedo through a hole in the top, 0.6 metre in diameter. About two or three tonnes of water from a leak also entered the torpedo and rested on top of the molten iron, separated from it by a thin crust of iron and slag. When the torpedo was moved, the water came into direct contact with the molten iron and vapourized with explosive speed and violence. Ninety tonnes of molten iron were blown out and the pouring spot, weighing over a tonne, was blown onto the roof of the building.

A flow of water onto the top of molten iron in an old-fashioned open ladle did not matter, as when it turned to steam there was plenty of room for the steam to escape. No-one realized, when the design of the ladle was changed, that entry of water would now be very dangerous.

The official report[2] described the cause of the water leak in detail – for 20 years plugs in the cooling system had been made of steel instead of brass and the report includes many photographs of corroded plugs – but this is irrelevant. If anyone had realized that water was dangerous, the flow of water from the leak could have been dammed or diverted.

Both the official report and the company report, which it quotes, made a series of recommendation designed to prevent water leaks. They also recommended that torpedoes should not be moved until any water in them has evaporated and that better protective clothing should be provided. The underlying cause of the explosion, however, was the failure to see the result of technical change. The weakness in the management system was the lack of any formal or systematic procedure for examining technical changes in order to see if there were any side-effects. The same weakness was present at Flixborough (Chapter 8) and in the accident discussed in Chapter 7.

The official report does quote the Safety Policy statement of the organization which required 'a progressive identification of all hazards involving injury and/or damage potential'. The official report concluded that 'senior management had not implemented the declared safety policy' and that they 'should take urgent and comprehensive action to implement it'. However, this advice is not very helpful. How do we identify all hazards? No advice was given.

For major modifications, hazard and operability studies, referred to several times already in this book, are now widely recommended. Simpler techniques have been described for minor modifications[3]. Both hazard and operability studies and the simpler techniques have been developed with the process industries in mind and may require some modification before they can be applied to industries such as steel production.

An explosion in a pumping station

In 1984 there was an explosion in a water pumping station at Abbeystead, Lancashire, which killed 16 people, most of them local people who were visiting the station. Water was pumped from one river to another through a tunnel. When pumping was stopped some water was allowed to drain out of the tunnel and leave a void. Methane from the rocks below accumulated in the void and, when pumping was restarted, was pushed through vent valves into a pumphouse, where it exploded.

If the presence of methane had been suspected, or even considered possible, it would have been easy to prevent the explosion by keeping the tunnel full of water or by discharging the gas from the vent valves into the open air. In addition, smoking, the probable source of ignition, could have been prohibited in the pumping station (though we should not rely on this alone; as mentioned above and in Chapter 4 we should always try to prevent formation of a flammable mixture). None of these things were done because no-one realized that methane might be present. Although there were references to dissolved methane in water supply systems in published papers, they were not known to engineers concerned with water supply schemes.

The official report[4] recommends that the hazards of methane in water supplies should be more widely known but this will prevent the last accident rather than the next. Many more accidents have occurred because information on the hazards, though well-known to some people, was not known to those concerned. The knowledge was in the wrong place. Another example was described in Chapter 4. Hazard and operability studies (hazops) will prevent these accidents only if a member of the hazop team knows of the hazard. Perhaps engineers should be encouraged to read widely and acquire a rag-bag of bits of knowledge that might come in useful at some time. At Abbeystead, if a hazop had been carried out, and if one member of the team had known, however vaguely, that methane might be present, and had expressed his misgivings, then the accident could have been prevented. It was not necessary for him to know the precise degree of risk or exactly what action to take; it would be sufficient for him to suspect that methane might be present and alert his colleagues.

Another significant point is not considered in the official report. An underground pumphouse is a confined space and many people have been overcome in such spaces. In particular carbon dioxide has often accumulated in underground confined spaces, with fatal results. Confined spaces in which there is any reason to suspect the presence of hazardous gas or vapour should be entered only after testing of the atmosphere, formal consideration of the hazards, gas-freeing and isolation (if necessary) and authorization by a competent person. In the UK this is required by the Factories Act, Section 30.

References

1. KLETZ, T. A. *Loss Prevention,* Vol. 9, 1975, p. 65.
2. Health and Safety Executive, *The Explosion at the Appleby–Frodingham Steelworks, Scunthorpe on 4 November 1975,* Her Majesty's Stationery Office, London, 1976.
3. KLETZ, T. A. *Chemical Engineering Progress,* Vol. 72, No. 11, Nov. 1976, p. 48.
4. Health and Safety Executive, *The Abbeystead Explosion,* Her Majesty's Stationery Office, London, 1985.

Chapter 15

Three weeks in a works

The accidents described so far in this book have been serious or unusual, usually both. By concentrating on these have we been giving a wrong impression and making recommendations that will have little impact on the majority of accidents?

To try to answer this question I have, on several occasions, with the help of colleagues, investigated every accident that occurred in a works during a three week period. This chapter describes the results of one such investigation. The accidents are described first and then some general observations are made. We shall see that, as with the major incidents described so far, we can learn much more if we look beyond the immediate technical recommendations for ways of avoiding the hazards or for ways of improving the management system.

During the three weeks we devoted more time to the investigation of the accidents than might be practicable in the ordinary way. Nevertheless the results showed that the investigations normally carried out were often superficial and that there was more to be learnt from the accidents than was in fact being learnt. In addition, conclusions could be drawn from the accidents as a whole that could not be drawn from individual incidents. Information bought at a high price was not being used.

The works employed about a thousand people, including maintenance workers; apart from a few clerical and laboratory workers, they were all male. Half the works consisted of large continuous plants making a few chemicals in large tonnages, while the other half consisted of batch plants making a range of small-tonnage products, many of which were corrosive.

An accident was defined as any incident that resulted in attendance at the ambulance room for first-aid treatment. Many of the accidents were therefore very trivial, e.g. dust in eye. Only one accident resulted in lost time. We also investigated dangerous occurrences, i.e. incidents that caused damage to plant or loss of product or raw material. Altogether there were 32 injuries and five dangerous occurrences (one of which caused injury) during the three weeks. We visited and photographed the scene of every accident and spoke to the injured men and to other people who were involved, and we obtained copies of the accident reports. A handwritten report was prepared by the foreman and completed by the manager for every incident. In addition, for the more serious incidents, including all chemical burns and all dangerous occurrences, a more extensive typed report was produced and circulated to the senior management and the company safety adviser. These typed reports were models of their kind; the descriptions were crisp but no relevant details were omitted and clear recommendations for action were made and accepted. In contrast the handwritten reports were often superficial and some of them did not arrive at the safety foreman's office until weeks later. A few never arrived at all, despite prompting by the safety foreman, who had the unenviable job of trying to extract them from foremen and managers who did not see the need to fill in a form every time someone got dust in his eyes. However, as we shall see, there is something to be learnt from almost every incident.

Nine of the accidents were due to dirt or liquid entering the eye. Most of them occurred in circumstances where men would not normally be expected to wear goggles. The only effective method of prevention would be for every employee to wear safety spectacles at all times, but the management of the works did not feel that the effort required to persuade everyone to do so could be justified. An attempt had been made, with some success, to persuade workshop employees to wear safety spectacles, and for some years all laboratory workers had had to wear them. As they were staff employees, it was easier to make rules and enforce them.

Although there are a few criticisms of the works in the following pages, nevertheless its accident record was good and on the whole the attitude to safety was above average. At the time of the survey the works was well on the way to completing a million hours without a lost-time accident.

The accidents

The incidents are described in the order in which they occurred.

(1) A process operator on one of the batch plants brushed his forehead with his arm and some dirt from his overalls entered his eyes.

According to the report he was told to take more care, a piece of accident prevention liturgy that we will meet again and again. The report also recommended that after carrying out dirty jobs, people should change into clean overalls. It did not discuss the practicality of this proposal. How often would they have to change? Are the overalls available? It might have been better to have looked round the plant to see if overalls generally were dirty. In general, when investigating accidents we need to know whether we have uncovered an isolated incident or if there is an ongoing unsatisfactory state of affairs.

(2) Half-an-hour later the *same man* dipped a small tank manually, using a dipstick. His goggles steamed up so he lifted them up to read the dip and then forgot to replace them. As he was replacing the cover on the dip-hatch a splash of liquid entered his eye. He was again told to take more care.

Eye injuries were discussed above. The liquid in the tank was very viscous but even so it ought to be possible to provide a level measuring instrument and thus *avoid* the situation that lead to the accident.

(3) A laboratory assistant complained of sore eyes. He may have rubbed them with contaminated hands, though he normally wears gloves. As the cause was uncertain, no recommendations could be made.

(4) The first dangerous occurrence: liquid was pumped intermittently from an atmospheric-pressure storage tank to one which was kept at a gauge pressure of 1 bar. The pump suction and delivery valves were kept open and a non-return valve was relied on to prevent back-flow (Figure 15.1). A piece of wire, about 3 inches (75 mm) long, became stuck in the non-return valve and the atmospheric-pressure storage tank overflowed. Two tonnes of liquid were spilt.

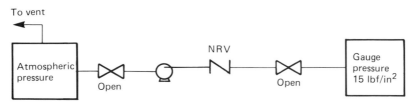

Figure 15.1   When the NRV failed, a leak was inevitable (see item 4)

The operators claimed that they had used this method of operation for seven years without incident. However, non-return valves are always liable to fail and a spillage in the end was almost inevitable. The incident is a classic example of an accident waiting to happen.

Note that seven years experience without a spillage does not prove that the method used is safe. All it tells us is that we can be 86% confident that the chance of a spillage is once in 3½ years, or less[1].

(5) While a welder was burning a hole in a pipe, 6 inches (150 mm) diameter, with walls ½ inch thick, in a workshop, a sudden noise made him jerk, his gun touched the pool of molten metal and a splash of metal hit him on the forehead. If the wall thickness had been above ½ inch, a hole would have been drilled in the pipe first, and the foreman said, on the accident report, that this technique should be used in future for all pipes. However, this would be troublesome. A crane would be needed to move the pipe to the drilling bay and back and this could cause delay. In practice, nothing was done.

The welder had cut thousands of holes before without incident, the chance of injury was small and it may have been reasonable to do nothing and accept the slight risk of another incident. In practice this is that the foreman, and the engineer in charge, decided to do. However, they were not willing to say so and instead they recommended action that they had no intention of enforcing.

After an accident many people feel that they have to recommend something even though they know it will be difficult, and they have no intention of carrying out the recommendation with any thoroughness. It would be better to be honest and say that the chance of another incident is so low that the risk should be accepted. The law does not expect us to do everything possible to prevent accidents, only what is 'reasonably practicable'. Whether or not that was the right decision in the present case is a matter of opinion.

(6) A small leak of liquefied petroleum gas (LPG) occurred from a drain valve on a pump suction line. It was detected by combustible gas detectors

which were permanently installed nearby and was stopped by closing the valve before the gas ignited.

The drain valve was far below the company standards. It was a brass valve, of a type stocked for use in domestic water systems, was connected by screwed joints, was inadequately supported, and was a single valve though the company standard called for two valves, 1 metre apart (or one valve and a blank if the drain point is used only occasionally). The works had spent a substantial sum of money upgrading the standard of its LPG installations and the programme had been completed less than a year before; it is not known how the substandard valve came to be introduced. The incident showed up weaknesses in the system for controlling plant modifications. It was far too easy for any foreman who wanted to do so, to alter the plant.

A foreman entered the cloud of vapour (about 2 metres across) to close the valve. Fortunately it did not ignite or he would have been seriously hurt. He should not have entered unless protected by water sprays. Perhaps practice in closing valves in this way should be included in fire training.

This incident resulted in three separate actions at company level:

(a) Improvement in the methods for controlling plant modifications. Other incidents, including the Flixborough explosion (Chapter 8) which occurred soon afterwards, emphasized the need for such improvements.

(b) An effort to discourage people from entering clouds of leaking gas to isolate leaks. I would not go so far as to say that no-one should ever do so; there may be occasions when, by taking a chance for a few moments, someone has prevented a leak developing and ultimately exploding, but such cases should be exceptional. We should try to avoid putting people in situations where they have to make such decisions. Remotely operated emergency isolation valves should be installed on the lines leading to equipment which experience shows is liable to leak. (See Chapter 4).

(c) A survey of other plants to see if there were any more substandard drain valves. Several were found.

If I had not been involved in the 'Three Weeks in a Works' exercise I might not have realized the significance of this incident and recommended the three actions above.

(7) During fog a motorcyclist hit a curb and came off. Though he was on company land the accident did not count towards the works record, as he was on his way home, and for this reason very little interest was taken in it, though it was the most serious accident to occur during the whole three weeks.

(8) A line had to be drained into a drum. The end of the line was 0.5 metre above the drum so not surprisingly the operator was splashed on the face with corrosive liquid. Fortunately he was wearing goggles.

A few years earlier there had been a number of accidents during sampling, and the need to locate the sample bottle immediately below the sample point had been emphasized in operator training. No-one recognized that the operation being carried out was similar to sampling and that the same precautions should be taken.

Other failures to recognize the true nature of a situation were described in Chapter 1 (no-one realized that a heavy lid under which a man was working was lifting gear) and in Chapter 7 (no-one realized that an open vent is a relief valve).

(9) While three men were lifting a long bent pipe (20 feet by 3 inches bore; 6 metres by 75 mm) onto a trailer, one man let go too soon and trapped another man's finger, breaking a bone. He did not lose any time.

Could men be persuaded in cases like this to lift the pipe with straps rather than bare hands?

(10) A ¼ inch nipple on an instrument line was screwed into a ¼ inch to ⅜ inch reducing bush (Figure 15.2). The nipple was leaking. While an artificer was trying to tighten it the bush broke and he was splashed in the face with a corrosive liquid. Fortunately he was wearing goggles.

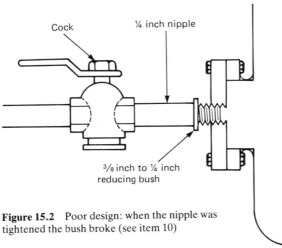

Figure 15.2   Poor design: when the nipple was tightened the bush broke (see item 10)

The plant had been built about 10 years earlier and the connection was below the current standards, which required robust flanged or welded connections between process equipment and the first isolation valve. Screwed connections of the type which broke would not have been permitted anywhere, even on instrument lines, for corrosive materials, and only after the first isolation valve for other materials. How far should we go in bringing old plants up to modern standards? Some changes are impossible. Replacement of any connections similar to the one that broke is so cheap and easy that it should certainly be carried out. In other cases we have to balance the seriousness of the risk against the cost of improvement[2].

Perhaps we should list all equipment which is not up to modern standards and decide how far to go in removing it. (See Chapter 17).

(11) A laboratory assistant knocked her elbow. She did not know where or when so no recommendations could be made.

(12) A welder lifted up his helmet in order to chip, did not put on ordinary goggles and a piece of slag entered his eye. Similar incidents had occurred

before and the welders had been supplied with some helmets in which the blue glass is hinged and there is white glass underneath. When a welder wants to chip he lifts up the blue glass. However, the welders never used these helmets. When asked why, they said that the windows were smaller than those in the ordinary helmets. This should not be difficult to put right but people find it easier to fling protective equipment aside as unsuitable rather than suggest improvements.

We noticed that most of the welders found their helmets too long and had cut about 2 inches (50 mm) off them. No-one had ever told the manufacturer.

(13) A fitter burnt his wrist on a small, hot, uninsulated pipe. As he stretched his arm he exposed his wrist. The accident could be prevented by a campaign to persuade fitters to wear gloves, but it is not certain that the risk justifies the effort required. Perhaps it is sufficient to publicize the risk and remind fitters that gloves are available. The incident would have been a suitable one for 'tool-box talks' had they been the practice on the works.

(14) A small hydrogen fire occurred in a small open pit in which some residues were dissolved in acid. The flames were lazy, and confined to the pit, and there was no hazard. The fire was soon extinguished with foam. Nevertheless the question was asked, 'Should we continue this operation in the open or would it be safer to change to a closed vessel?'

It was then realized that if this was done the vessel would have to be inerted with nitrogen and fitted with an oxygen alarm; a level measuring instrument would be required and numerous other complications, which were expensive and might go wrong. It was simpler and safer to continue with the present method, but a foam generator was permanently installed nearby[3].

(15) A craft assistant sprained a finger while helping a fitter to pack a gland on a reciprocating pump. The job was being done by an incorrect but quick method which gives poor results but is often tolerated by foremen and engineers. Could a safe and efficient but quicker method have been found? If not, the foremen should have explained why the correct method should have been used and then have enforced its use.

(16) Before moving some liquid from one tank to another an operator failed to check the position of all the valves; a valve leading to a third tank had been left open and this tank was overfilled. Two tonnes of liquid were spilt. The valve had been left open because the line, a long one, was not fitted with thermal relief valves.

It is easy to blame the operator but the mistake was one that it is easy to make. If thermal relief is needed a relief valve should be fitted. We should not in a case such as this rely on operators who may forget to open (or close) a valve.

(17) While making a thermocouple an artificer caught his thumb on the sharp end of the insulation. A tool could be devised so that the operator's thumb is not in direct contact with the insulation but the operation is carried out infrequently and it is therefore doubtful if the effort involved would be justified or if the tool would be found when required.

(18) An electrician was running a temporary cable on a construction site when he stumbled on a piece of wire mesh, lost his balance and hurt his foot on a metal bar.

The foreman told the man to be more careful when crossing ground disturbed by contractors. The engineer endorsed the comment. Our first reaction was to agree. However, when we visited the site we found that the area where the electrician was working had not been disturbed by the contractors. It was covered by weeds, 2 feet (60 cm) tall, which hid numerous tripping hazards, many sharp and sticking up at all angles. The ground should have been bulldozed by the contractors before work started.

An accident report may seem plausible when read in the office but a visit to the site may tell a different story. An accident report should never be completed before the site has been seen. (See accident 33).

This is the only accident due to untidiness, though in UK industry as a whole it is one of the biggest causes of accidents, the biggest according to some reports.

(19) While picking up some bolts from above eye level a man dislodged some rust which entered his eye. This is another of those eye injuries which could have been prevented only by the wearing of safety spectacles at all times.

(20) A laboratory assistant was wearing thin plastic gloves while handling a corrosive chemical. The gloves split and he burnt his hand. It was agreed that thicker gloves should be used in future.

(21) A choke occurred in a steam-traced line. The insulation was removed and while a man was fixing up a steam lance, to clear the choke, he burnt his hand on the steam tracing. The steam tracing was inadequate and chokes were common. The most effective method of prevention was better tracing, not easy on an old plant but something to note for new plants.

(22) While a process operator was using a wheel-dog to open a valve it moved suddenly and trapped his finger against a bracket. The bracket served no useful purpose and was removed.

(23) Dust in eye. See comments on accident 19.

(24) A man slipped and fell in an area which was known to be very slippery. The manager commented, 'The need to be more careful will be stressed once again'. The manager, the foreman and the injured man all agreed that nothing could be done and that occasional accidents were inevitable. In fact, it would have been easy to fit a non-slip coating on the wooden blocks on which the man slipped, or to fit studs to his boots. In UK coal mines, slipping, a major cause of accidents, has been greatly reduced by fitting anti-slip studs to boots[4].

(25) While climbing into the space below the roof of a floating roof tank, which was being repaired, a foreman bumped his head. His hard hat fell off and he hit his head on the steelwork. Afterwards it was found possible to improve the access.

(26) A man forgot to put on his safety spectacles before using a portable grinder and got a speck of dust in his eye. Though he blamed himself, the

oversight is an easy one to make. If he had worn spectacles at all times, the injury would not have occurred.

(27) An electrician caught his wrist on a metal fixing band. Plastic bands are cheaper and safer.

(28) A man was removing empty 40 gallon (180 litre) drums which were stacked four-high on their sides (Figure 15.3). The method used was to chock the tier next to the end, remove the end chock and allow four drums to roll down. One drum rolled to one side and as the area was congested the man could not move out of the way. The accident report said, 'To be extra careful when de-stacking drums'. In fact, the only effective method of prevention was to remove the congestion; not easy, as space is limited.

(29) Full 40 gallon drums were being loaded onto a flat wagon with a forklift truck. The drums were stored on their sides in tiers, as in Figure 15.3. Two drums were lifted onto the wagon at a time (Figure 15.4) and then tipped into a vertical position with the ends of the forks (Figure 15.5). One tipped too far and fell onto the foot of the driver who was standing on the wagon to adjust the position of the drums.

Figure 15.3   Method used for storing drums (see item 28)

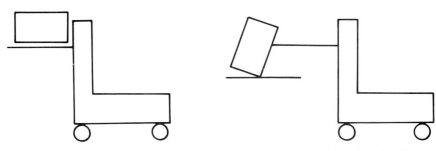

Figure 15.4   Drums are lifted two at a time like this onto a flat wagon . . .

Figure 15.5   . . . and then tipped up like this (see item 29)

Several similar accidents had occurred before. Despite several investigations no better method had been devised and the problem had been put aside as unsolvable.

One possible solution would have been to store the drums on their ends and lift them onto the wagons with a suitable attachment fitted to the forklift truck. However, this would need extra storage space and would not be the complete answer, as a layer of horizontal drums was put on top of the vertical layer (Figure 15.6).

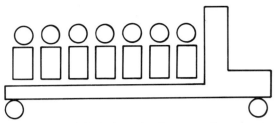

Figure 15.6   Method of carrying drums on a flat wagon (see item 29)

One customer would not accept wagons with the upper horizontal layer of drums as a man had been killed when one of these drums fell off. This illustrates one of the problems of industrial safety. There may be a case for forbidding the top layer of drums at all times; there may be a case for allowing it to continue. There is no logic in stopping it at the site where the accident occurred but allowing it to continue elsewhere.

We felt that this problem called for more expertise than we, or any of the local management, possessed and we recommended that it was referred to consultants.

(30) A craft assistant cut his finger on a ragged thread while cleaning the threads of a flange with a cloth and cleaning fluid. Safety apart, cloths should not be used for cleaning threads, as bits of the cloth get left behind. A brush should be used.

(31) While hoisting a piece of pipe onto a trailer, a speck of dust fell into a man's eye. He was wearing sight-correcting spectacles, so safety spectacles would not have helped, unless fitted with sidepieces.

(32) An operator slipped on the deck of a ship (the works had its own jetty). As with accident 24, studded footware might have prevented this accident but in this case we had to balance the decreased risk of slipping with the slightly greater risk that the studs would cause a spark on a concrete floor and ignite any flammable vapour which was present. The latter risk was very slight.

(33) A fitter was fixing a clip onto a hose using a clamping machine. It had been mounted parallel to the edge of the bench instead of at right angles, so that the fitter caught his finger on an object on the bench. The accident report did not say this. As with accident 18, a visit to the site of the accident told us much more than the report.

(34) A man slipped while entering the works, put his hand out to steady himself and caught it on a sharp protrusion on the gate. The protrusion, and another one, were removed. The gate is never closed so there is probably no need for it to be there.

(35) Dust in eye, while using an impact wrench out-of-doors on a windy day. See accidents 19 and 23.

(36) A planner slipped while inspecting a construction site. See accidents 24 and 32.

General comments

As with the serious accidents discussed in earlier chapters, these trivial accidents can teach us far more than is apparent at first sight. The local managers usually failed to see the inner layers of the onion, partly because they did not realize that there were inner layers waiting to be seen but also because they did not consider that trivial injuries were worth much of their time or attention. As already stated, only one accident, involving a motorcycle, caused absence from work. However, many more could have caused more serious injury but by good fortune did not do so.

The most important general lessons to come out of the study were:

(1) If we believe that the accident is so unlikely to happen again that we should do nothing, we should be honest and say so, and not recommend actions that we have no intention of carrying out. (See accident 5.) (Of course, after a serious accident we may have to do more than we think is technically necessary, in order to reassure those exposed.)

(2) A procedure is needed for the control of plant modifications. They should not be made unless authorized by a responsible member of management, who should first try to identify the consequences and should also specify the standard to be followed. (See accident 6 and Chapters 7 and 8.)

(3) Men should not enter clouds of flammable gas unless protected by water spray. (See accident 6 and Chapter 4.)

(4) We should carry out periodic inspections to identify substandard equipment (see accidents 6, 10 and 16) and procedures (see accidents 4, 8, 15, 28, and 29). These inspections can be carried out in two ways:
  (a) A walk round looking for anything substandard. It is a good plan to take a camera loaded with slide film with you and photograph the hazards that you see. They can then be shown to those who work in the area. People are shocked to see the hazards that they have passed every day without noticing.
  (b) However, hazards are more likely to be found if we look at specific items of equipment such as drain points or screwed joints, or specific practices such as sampling or drum handling. Recent incidents, on our own works and elsewhere, and a glance at old reports, will provide subjects for inspection.

(5) Visit the scene of an accident. We will get misleading impressions if we rely solely on written reports. (See accidents 18 and 33.)

(6) If protective equipment is not being used, ask why. (See accident 12.)

(7) Reports which simply suggest that someone should take more care should be investigated further. There is usually something that managers can do to reduce the opportunities for human error or to protect against the consequences. (See accidents 1, 2, 18, 24 and 28, and reference 5.)

(8) Do not rely on procedures when there is a cheap and simple way of removing a hazard by modifying the plant. (See accident 16.) Safety by design should always be our aim. Very often there is no 'reasonably

practicable' design solution, as these incidents show, and we have to depend on procedures; but if there is one we should make use of it.

Analysis of the accidents suggests that:

Twelve accidents, 33% of the total (6, 7, 8, 10, 14, 16, 17, 21, 22, 25, 28 and 34), could have been prevented by better design or layout; the changes required were mostly minor.

Seven accidents, 20% of the total (4, 5, 9, 15, 29, 30 and 33), could have been prevented by better methods of working.

Thirteen accidents, 36% of the total (1, 2, 12, 13, 19, 20, 23, 24, 26, 31, 32, 35 and 36), could have been prevented by better protective clothing. The works handles a lot of corrosive chemicals and the figure might be lower elsewhere. On the other hand this group includes a number of 'dust in eye' incidents which could have been prevented by the wearing of spectacles.

One accident, 3% of the total (18), could have been prevented by better tidiness.

Insufficient information was available for three of the accidents, 8% of the total (3, 11 and 27).

(9) People need to be reminded of the limitations of experience. Twenty years without an accident does not prove that the operation is safe unless an accident in the 21st year is acceptable. To be precise, 20 years experience does not even prove that the average accident rate is less than once in 20 years. All it tells us is that we can be 86% confident that the average rate is less than once in 10 years. (See accident 4 and reference 1.)

Conclusions

The incidents described in this chapter have confirmed the theme of this book, that there is much more to be learnt from accidents than we usually learn, not because we are not aware of the facts but because we do not consider them deeply enough. In particular, valuable lessons can be learnt from trivial accidents. (See also Chapter 1.) In addition, two more general conclusions may be drawn:

(1) Too much writing and talking on safety is concerned with generalities. We can learn much more if we discuss specific incidents. No-one should ever generalize without at least describing some incidents which illustrate and support his case.

(2) To quote Grimaldi[6]:

> '(Improvement in safety) is more certain when managers apply the same rigorous and positive administrative persuasiveness that under-lies success in any business function . . . outstanding safety performances occur when management does its job well. A low accident rate, like efficient production, is an implicit consequence of managerial control'.

A similar view was expressed 20 years ago following an enquiry into factory accidents by two factory inspectors[7]:

'It also produced evidence to confirm that established and generally accepted methods of accident prevention did succeed; and several impressive examples were found of improvements achieved by the energetic and diligent application of principles which had long been advocated, but which had not been put into practice earlier with sufficient thoroughness'.

References

1. KLETZ, T. A. *Myths of the Chemical Industry,* Institution of Chemical Engineers, Rugby, UK, 1984, p. 80.
2. KLETZ, T. A. *Loss Prevention,* Vol. 14, 1981, p. 165.
3. KLETZ, T. A. *Cheaper, Safer Plants,* Institution of Chemical Engineers, Rugby, UK, 2nd edition, 1985, p. 109.
4. *Tungsten Carbide Tipped Studs for Anti-Slip Boots,* Safety in Mines Research Establishment, Sheffield, UK, 1971.
5. KLETZ, T. A. *An Engineer's View of Human Error,* Institution of Chemical Engineers, Rugby, UK, 1985.
6. GRIMALDI, J. V. *Management and Industrial Safety Achievement,* Information Sheet No. 13, International Occupational Health and Safety Information Centre (CIS), Geneva, Switzerland, 1966.
7. *Employment and Productivity Gazette,* Oct. 1968, p. 827.

Appendix: An example of a well-written accident report

Reports such as the following were produced for all the dangerous occurrences, for all the chemical burns and for those other accidents where injuries might easily have been more serious, e.g. accidents 28, 29 and 32. Handwritten reports were produced for the other accidents. This report deals with accident 10. It is not the original one.

Minor accident report No. . . .

Date and time	17 June 15.00 hrs.
Place	E304 still, ground floor.
Injured person	Mr B. Black, instrument artificer, days.
Injuries	Chemical burns to face.
Date of investigation	18 June.
Investigation team	Mr B. Brown, plant manager
	Mr G. Green, process foreman
	Mr G. Grey, maintenance foreman
	Mr W. White, safety representative
	Mr S. Smith, safety officer
	Mr J. Jones, process operator (witness)
	Mr B. Black

What happened?   At about 14.00 hours on 17 June an operator told Mr Green that there was a slight leak from a screwed joint on the base on E304 still. After inspecting the leak Mr Green asked Mr Grey to repair it and issued a permit-to-work for the job. It stated that the leaking equipment contained corrosive materials . . . under a hydrostatic pressure of about 15 feet (0.5 bar) and that gloves and goggles should be worn. The permit was produced and was found to be in order. Mr Grey accepted the permit and asked Mr Black, an experienced craftsman, to tighten the joint. He showed him the permit.

The joint is shown in the attached diagram. [See Figure 15.2.] The leak was coming from the screwed joint between the ¼ inch nipple and the reducing bush. The liquid was trickling out, not spraying. Mr Black attempted to screw the nipple further into the bush but while he was doing so the bush broke and a stream of liquid came out. Mr Brown was standing to one side, having foreseen the possibility of the bush breaking, and the main stream of liquid passed by him but some spray came in contact with his face. The joint was at chest height, Mr Black was wearing goggles.

Mr Black immediately put his face under a shower which was located about 10 feet [3 metres] away. He was assisted by Mr J. Jones, process operator, who happened to be passing. Mr Jones sent for the ambulance and Mr Black was taken to the Medical Centre but was not detained. There is some discoloration on his face which is expected to last for a few days.

Discussion   The joint which broke was installed about 10 years ago when the plant was built. It is not up to the current standard, which states that screwed connections may be used only on narrow-bore instrument lines, only after the first isolation valves, and only if the equipment contains non-hazardous materials (i.e. does not contain corrosive liquids, toxic liquids or gases or flashing flammable liquids). (See Engineering Dept. Standard No. . . . .)

While it is clearly impossible to bring all old equipment up to modern standards, the joint which leaked is so far below them that any other similar equipment should be replaced as soon as possible.

Actions   Mr Brown will arrange for one of the process foremen to be released from his normal duties for a day, or longer if necessary, so that he can survey the plant and list any other substandard joints. These will be replaced during the shut-down scheduled for September.

ACTION: Mr B. Brown
Mr G. Grey

The safety officer will ask other plants that handle corrosive or other hazardous materials to carry out similar surveys.

ACTION: Mr S. Smith

The Company Safety Adviser will be asked to inform other factories.

ACTION: Company Safety Adviser

Completion date   See above. This report should be brought forward on 1 September to confirm that substandard joints have been listed and are on the shutdown list and again after the shutdown to confirm that they have been replaced.

ACTION: Works office

Resources required   This depends on the number of substandard joints found. A quick look suggests that there could easily be 10–20, perhaps more. If so the cost of replacing them will be about £. . . . This will not affect the length of the shutdown.

B. Brown
Plant manager

Circulation

Works manager   Works engineer
Plant engineer   Those present at enquiry
Engineering Dept   Company safety officer
Plant notice board   Works office
Medical officer
Personnel officer (for file of injured man)
Managers of plants with similar hazards

Note added by works manager

There must be many other items of equipment on the works which are below current standards. How far should we go in replacing them? I am writing to the Company Safety Adviser asking him to review this question, with Engineering Dept, and prepare a note which will list the types of equipment involved and the extent of the problem and make recommendations.

What would have happened if Mr Jones had not been passing? In some factories, operation of a shower sounds an alarm in the control room. Should we adopt this system? What will it cost? Will the safety officer please consider and make recommendations.

I shall also ask the Company Safety Adviser to send a copy of this report, suitably edited, to the Institution of Chemical Engineers Loss Prevention Bulletin.

B. Bigshot

Chapter 16

Some pipe failures

This chapter discusses a number of pipe failures. The recommendations made for preventing further failures are important in themselves and also show that by looking at a group of related accidents we may be able to draw conclusions which we could not draw from single incidents.

Why are pipe failures important?

Since Flixborough (Chapter 8) there has been an explosion of papers on the probability of leaks and on the behaviour of the leaking material – how it disperses in the atmosphere, what pressure is developed if it ignites etc. – and expensive experiments have been carried out. In contrast, little attention has been paid to the reasons why leaks occur and to ways of preventing them. It is as if leaks were inevitable. Yet if we could prevent leaks we would not need to worry so much about their behaviour.

Table 16.1, derived from a review of 67 major leaks of flammable gas or vapour[1], most but not all of which ignited, shows that half were the results of failures of pipes or pipe fittings, over half if we ignore transport accidents. If we want to prevent major leaks, the most effective action we can take is to prevent pipe failures.

Table 16.1 Origin of major leaks of flammable gases and vapours

Origin	Number of incidents	Notes
Transport container	10	Includes 1 zeppelin
Pipeline (including valve, flange, bellows, sight-glass, etc.)	34	Includes 1 sight-glass and 2 flexes
Pump	2	
Vessel	5	Includes 1 internal explosion, 1 foam-over and 1 case of overheating
Relief valve or vent	8	
Drain valve	4	
Maintenance error	2	
Unknown	2	
Total	67	

There is not sufficient detail in reference 1 to analyse the reasons for the 34 pipe failures, so instead I have analysed about 50 pipe failures (or near-failures) on which I have seen reports, most of them unpublished. Many of the failures did not result in fires or explosions, as the leaks failed to ignite, the contents were non-flammable or the leak was prevented at the last minute. The incidents are summarized below.

A – Some pipe failures that could have been prevented by better design

In some cases the piping designer was not provided with the information needed for adequate design.

(A1) Water was added to an oil stream using a simple T-junction (Figure 16.1). The water did not mix with the oil and extensive corrosion occurred downstream of the junction. There was a major fire.

The corrosion would not have occurred if the water pressure had been higher, so that the water hit the opposite wall of the pipe, or if the water had been added to the centre of the oil line as shown in Figure 16.2.

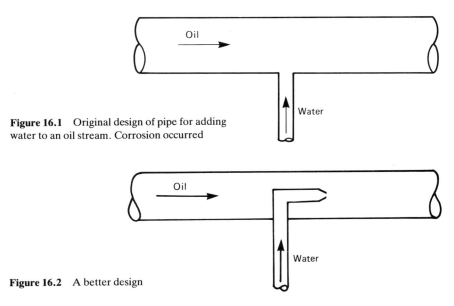

Figure 16.1   Original design of pipe for adding water to an oil stream. Corrosion occurred

Figure 16.2   A better design

(A2) A device to achieve this was designed for another plant. It was assembled as shown in Figure 16.3. Corrosion was worse! Once the device was impossible to tell if it had been assembled correctly. It should have been designed so that there was a telltale sign on it or, better still, so that it could not be assembled incorrectly.

(A3) A 4 inch (100 mm) branch was fitted onto an 8 inch (200 mm) pipe at an angle of 45° (Figure 16.4). The line failed and 30 tonnes of hot, flammable hydrocarbon were released; fortunately it did not ignite.

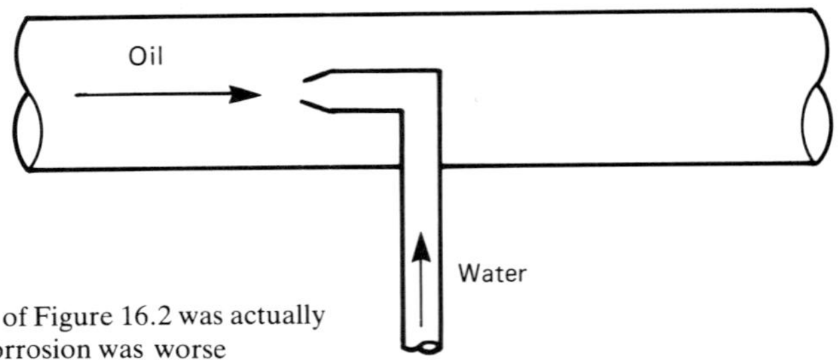

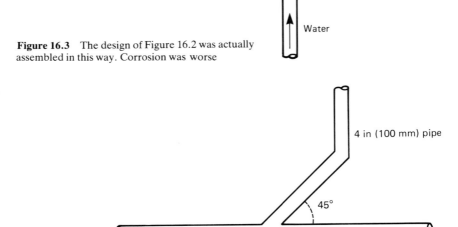

Figure 16.3   The design of Figure 16.2 was actually assembled in this way. Corrosion was worse

Figure 16.4   A poorly designed joint which failed, causing a major leak

Failure was the result of vibration and fatigue, enhanced by the poor detailed design of the joint, the different wall thicknesses of the two pipes and the oval shape of the hole. In addition the smaller line should have been better supported and when vibration occurred, the reason for it should have been investigated.

(A4) A valve was opened in error and cryogenic liquid came into contact with a mild steel pipe, which disintegrated.

We would never tolerate a situation in which operation of a valve at the wrong time could result in equipment being overpressured. We would install a relief valve. Similarly, we should not allow equipment to be overcooled by operation of a valve at the wrong time. The designer should have specified a grade of steel suitable for low temperatures or installed an interlock to prevent the valve being opened if the temperature is low.

This incident illustrates the very different attitudes adopted towards pressure and temperature. Too little or too much of either may be hazardous, and it is important to prevent either condition occurring[2].

(A5) Some wet hydrocarbon gases had to be blown down for a prolonged period. An ice-hydrate plug blocked an 18 inch (450 mm) blowdown line. It was cleared by external steaming. When the plug became loose the pressure above it caused it to move with such force that the line fractured at a T-junction.

The process designers did not foresee that wet gases might have to be blown down for so long. It should not occur very often. Nevertheless the incident would not have occurred if the design team had foreseen that prolonged blowdown might occur and had tried to prevent choking, e.g. by steam heating.

(A6) A stainless steel line intended for use with hydrogen at 360°C was fitted with a branch leading to a relief valve. There would normally be no flow through the branch and it was calculated that 1 metre along it the temperature would have fallen below 100°C and that mild steel could be used. However, the line failed because of hydrogen attack. Possibly the method used for calculating the temperature in the line did not allow for the shielding effect of surrounding pipes.

(A7) A bellows was found to be distorted. It had been designed for normal operation but the piping designer had not been told that it would be hotter when it was steamed out before a shutdown.

(A8) A line carrying heat transfer oil failed by fatigue as the result of repeated expansion and contraction as the temperature varied. The oil fell onto some cables and attacked them and a short circuit ignited the oil. There should have been more expansion bends in the line. (Flanges on oil lines should be located so that any drips do not fall on cables.)

(A9) Water froze in an LPG drain line and fractured a screwed joint.
Screwed joints are not suitable for process materials except for small-bore lines after the first isolation valve. (See Chapter 15, accident 10.) Flanged or welded joints should be specified.

(A10) A valve in a 10 inch (250 mm) liquefied butane line was located in a pit which was full of rainwater contaminated by sulphuric acid from a leaking underground line nearby. The bolts holding down the valve bonnet corroded and the bonnet flew off. A massive leak of butane exploded, killing seven people and causing extensive damage.
At first there was only a small leak from the valve. It was decided to empty the line by washing it out with water at a gauge pressure of 110 lbf/in^2 (7.5 bar), more than the usual operating pressure 50 lbf/in^2 (3.3 bar). The line was designed to withstand this pressure but in its corroded state it could not[3].
Although it was not good practice to install a valve in a pit that was liable to be flooded, it was also not good practice for the operating team to tolerate the flooding or to try to sweep out a corroded line using water at a higher pressure than that normally used.

(A11) A beryllium/copper circlip was used to secure the joints in an articulated arm carrying ammonia. The joint blew wide open. Stainless steel should have been specified. Afterwards it was decided to stock only stainless steel circlips so that errors could not occur.

(A12) The feed line to an ammonia plant failed suddenly as the result of hydrogen attack. The grade of steel used was considered suitable for the duty at the time the plant was built (about 1970), but since then the Nelson curves, which define the conditions under which hydrogen attack will occur, have been revised[4]. This is the only pipe failure I know of which occurred as a result of an error in a specification.

(A13) On a methanol plant, the joint between a 6 inch (150 mm) synthesis gas pipe and a flange failed and the escaping gas exploded. No one was hurt but damage was extensive. The report said that weld-neck flanges should be used for cyclic operations, as they have 10 times the fatigue life of the joint actually used, a combination of lap-joint and stub-end[5].

B – Some pipe failures that could have been prevented by better inspection during or after construction

(B1) A temporary support across an expansion loop was left in position. Fortunately it was spotted as the plant was warming up.

(B2) The exit pipe from a high-pressure converter was made from mild steel instead of ½% Mo. The pipe failed through hydrogen attack, a leak occurred at a bend and the reaction force pushed the converter over.

This was the most spectacular of many incidents that have occurred because the wrong material of construction was used. Whenever the use of the correct material is critical this should be indicated on the drawings, and all incoming steelwork (pipes, flanges, welding rods as well as fabricated items) should be checked before installation. Every piece of pipe, flange etc. should be checked, not just a sample.

Reduction in the number of grades of steel used will reduce the chances of error. It is a false economy to specify a cheaper grade for a limited number of applications if it increases the chance of error, during maintenance as well as construction. (See item A11.)

Six pipes in an ammonia plant were found to be cracked. The correct grade of steel had been used by it had received the wrong heat treatment[6].

(B3) A construction worker cut a hole in a pipeline at the wrong place. Discovering his error, he welded on a patch but said nothing. The line was radiographed but this weld was omitted as the radiographer did not know it was there. The line was then insulated.

The weld was substandard and leaked, and several men were gassed by phosgene, one seriously.

It is easy to talk of criminal negligence, but did the man understand the importance of good welding, or realize the nature of the materials that would be in the pipeline or what might happen to other people if his work was substandard? Unfortunately most construction engineers do not believe it is practicable to try to explain these things to construction workers.

(B4) Underground propane and oxygen lines leaked and an explosion occurred underground. The report said, 'During the construction of the pipework, doubts were expressed by the works management as to the quality of the workmanship and the qualifications of those workers employed'.

Why did they not do more than just express doubts to each other?

It is not good practice to run pipes underground in a factory, as the ground is often contaminated by chemicals which cause corrosion (although that did not occur in this case). If pipes, however, are run underground they should be wrapped, surrounded by clean sand or gravel and cathodically protected. When an unprotected line leaked, 140 tonnes of product were lost over a period of four days through a hole only ½ to ¼ inch (3–6 mm) diameter. As the leak occurred over a public holiday, the records office did not notice the discrepancy between the quantities despatched and received.

A leak in an underground line contributed to incident A10.

(B5) A portable hand-held compressed air grinder, in use on a new pipeline, was left resting on a liquefied petroleum gas line. The grinder was not switched off when the air compressor was shut down. When the air compressor was started up some of the line was ground away.

(B6) Several hangers failed and a new pipe sagged, over a length of 14 metres. When the pipe was installed, it did not have the required slope, so the contractors cut some of the hangers and welded them together again. One of the welds failed. Other hangers failed as a result of incorrect assembly and lack of lubrication.

After a 22 inch (550 mm) low-pressure steam main developed a crack it was found that the spring on one support was fully compressed, that a support shown on the drawing was not fitted, that another support was fitted but not attached to the pipe, and that the nuts on another support were slack.

(B7) Two lengths of 8 inch (200 mm) pipe which had to be welded together were not exactly the same diameter so the welder joined them with a step between them over part of the circumference (Figure 16.5). The pipe was then insulated. The botched job was not discovered until 10 years later, when the insulation was removed for inspection of the welds.

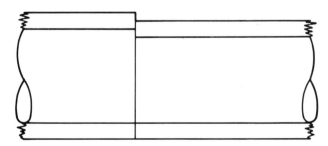

Figure 16.5   Two pipes of slightly different diameter were welded together

(B8) A bellows blew apart a few hours after installation. The split rings which supported the convolutions and equalized expansion were slack, with gaps of up to ½ inch between the butts of the half rings. The plant was not inspected before commissioning.

On another occasion a bellows was damaged before delivery and this was not noticed by the construction team.

Other bellows have failed because they were installed between two pipes which were not exactly in line and the bellows were allowed to take up the misalignment. If fixed piping does not meet exactly, it can be pushed until it does. If bellows are used, the piping must be made with more, not less, accuracy. Bellows should not be used to compensate for misalignment, unless specially designed for that purpose.

A near failure of a bellows was described in item A7. The most famous bellows failure of all time – Flixborough – was described in Chapter 8, where it was suggested that when hazardous materials are being handled, designers should avoid the use of bellows and use expansion loops instead. What you don't install, cannot be badly installed.

(B9) The supports shown in Figure 16.6 were bolted to the concrete bases and welded to the pipes. When the control valve closed suddenly, the shock caused the pipe to move suddenly and a piece was torn out of one of the bends. The escaping liquid caught fire and damage was extensive[7].

It is unlikely that the designer asked for such rigid fixing of the pipe. The details were probably left to the construction team.

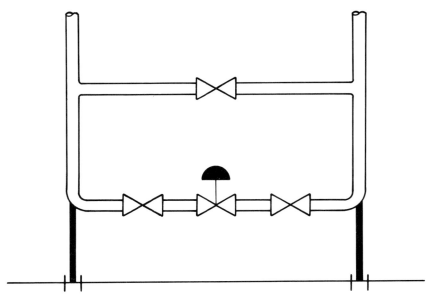

Figure 16.6   The supports were bolted to the concrete bases and welded to the pipes. When the pipe moved, a piece was torn out

(B10) A 10 inch (250 mm) pipe was fitted with a ¾ inch (20 mm) branch. The main pipe rested on a girder and there was a gap of 5 inches (125 mm) between the branch and the girder (Figure 16.7). When the pipe was brought into use its temperature rose to 300°C. The branch came into contact with the girder and was knocked off. Calculation showed that the pipe, which was 120 feet (36 metres) long, had moved 6 inches (150 mm).

Again, it is unlikely that the designer specified precisely how the branch was to be positioned in relation to the girder; it is probable that he left this detail to be decided on site. The construction workers would not have known the operating temperature of the pipe or how much it would expand.

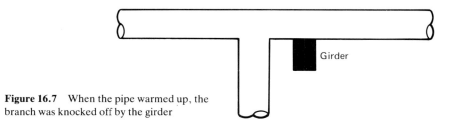

Figure 16.7 When the pipe warmed up, the branch was knocked off by the girder

(B11) An old pipe was re-used. Its condition was not checked and it failed as a result of corrosion/erosion which had occurred during its previous use.

On another occasion an old pipe that was re-used had used up much of its creep life and failed in service, where a new pipe would not have failed. A stream of high-pressure gas produced a flame 30 metres long. The pipe that failed had been used for 12 years at 500°C, but creep cannot be detected until a pipe is close to bursting.

(B12) A pipe was laid on the ground and corroded. The designer probably left the choice of support to the construction team.

(B13) Several pipes were inadequately supported, and so they vibrated and failed by fatigue. Supports for small-bore pipes are often not specified by designers but decided on site. It is often difficult to know whether or not a pipe will vibrate until the plant is on-line, and perhaps therefore the most effective way of preventing these incidents is for the operating team to take action when pipes are vibrating. Unfortunately they are often too busy during start-up and then the pipes become part of the scene and are not noticed.

(B14) A line carrying a material of 40°C melting point was kept hot by steam tracing. One section of tracing was isolated and the line set solid. Expansion of liquid in the rest of the line caused it to burst (Figure 16.8).

The detail of the steam tracing was probably not specified by the designer but left to the construction team. It should not have been possible to isolate the heating on part of the line.

(B15) A level controller fractured at a weld, producing a massive release of oil, hydrogen and hydrogen sulphide. The report said that the failure was due to poor workmanship and 'emphasises the need for clear instructions on all drawings and adequate inspection during manufacture'.

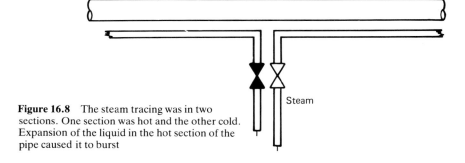

Figure 16.8 The steam tracing was in two sections. One section was hot and the other cold. Expansion of the liquid in the hot section of the pipe caused it to burst

(B16) A 1 inch (25 mm) screwed nipple blew out of a line carrying heavy oil at 350°C. Most of the plant was covered by an oil mist 100 feet (30 metres) deep. Fortunately it did not ignite.

The nipple had been installed 20 years earlier, during construction, for pressure testing, and was not shown on any drawing. If its existence had been known it would have been replaced by a welding plug.

(B17) A line carrying liquefied gas was protected by a small relief valve which discharged onto the ground. The ground was not level and after heavy rain the end of the tailpipe was below the surface of a puddle (Figure 16.9). The water froze and the line was overpressured.

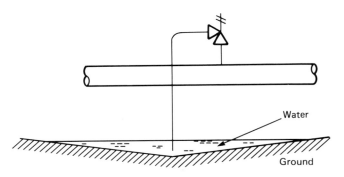

Figure 16.9   The end of the relief valve tail pipe was so close to the ground that it was below the surface of a puddle and was blocked when the puddle froze

(B18) Sometimes it is necessary to reinforce the wall of a pipe, where there are additional stresses from a branch or a support, by welding on a plate. There is usually a small gap, up to ¹⁄₁₆ inch (1.5 mm), between the plate and the wall of the pipe. This space should be vented, by a ¼ inch (6 mm) hole or a gap in the welding, or a pressure may develop when the pipe gets hot. The rise in pressure will be greater if any water is trapped in the space and turns to steam.

A steam pipe designed to operate at a gauge pressure of 200 lbf/in² (14 bar) collapsed because a reinforcing pad was not vented. A blowdown main collapsed for the same reason, fortunately during heat treatment.

(B19) Many pipes have failed because water collected in dead-ends and froze, or corrosive materials dissolved in the water and corroded the pipe. A 3 metre branch on a natural gas pipeline was corroded in this way; the escaping gas ignited and five men were killed.

Dead-ends are often the result of modifications but they are sometimes installed in new plants to make extension easier. They should point upwards, not downwards, so that water cannot collect in them. Most oil streams contain traces of water.

Dead-ends may be formed by removal of unused equipment (Figure 16.10) or simply by not using equipment. One of the two pumps shown in Figure 16.11 was rarely used. It was at a slightly lower height than the other pump so that water collected in the branch, which corroded right through.

Prevention of incidents such as these lies more in the hands of the operating team than of the constructors.

A piece of pipe was welded onto a liquefied petroleum gas line to support an instrument (Figure 16.12). Water collected in the support and corroded the LPG line. Gas was seen blowing out of the support!

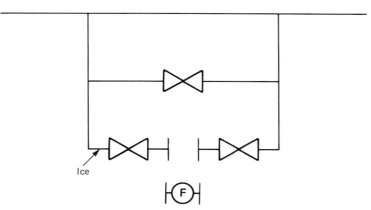

Figure 16.10   When the flowmeter was removed it formed a dead-end in which water collected

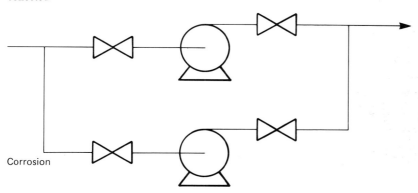

Figure 16.11   The lower pump was rarely used. Water collected and corrosion occurred

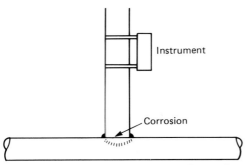

Figure 16.12   Water collected in a piece of pipe, used as an instrument support, and corrosion occurred

(B20) A leak occurred on a T-piece on a high-pressure boiler feed water main four years after start-up. The construction team had found that insufficient forged T-pieces were available, and had therefore made three by welding together short lengths of pipe. If they had told the operating team that they had done this then the substandard T-pieces could have been scheduled for replacement at the first opportunity. Keeping quiet about the bodging compounded the original error.

(B21) As this section has been a catalogue of construction failures, and of failure to spot the failures, let us end it with a success story. An alert welding inspector was looking through the radiographs of some pipework intended for use with refrigerated ammonia. He noticed that a series of radiographs said to have been taken from different butt welds were in fact actually taken on the same weld. The inspector had been engaged by the client, not the contractor[9].

C–Some pipe failures that could have been prevented by better operations

Some such failures have already been mentioned. (See items A10, A11, B13 and B19.) Here are a few more.

C1) A crane was used to move a line full of highly flammable liquid so that a joint could be made. A branch on the line was knocked off and a leak occurred.

(C2) After completing a lift, a crane had to be moved about 100 metres along a road for its next job. The driver decided not to bother lowering the jib and he hit and damaged an overhead pipeline.

If cranes frequently travel along a roadway, any pipes that cross can be protected by 'goalposts', girders across the road a few metres either side of the pipes.

(C3) An unforeseen decomposition occurred in a reactor, the temperature in the exit pipe rose rapidly and the pipe ruptured.

(C4) A little-used line was left full of water during the winter and was split by frost.

Conclusions

Some recommendations that arise out of a single incident or a few incidents have already been made. In this section we try to draw some conclusions from the incidents as a whole.

The classification of some of the incidents may be disputed but the general conclusion seems inescapable: the most effective action we can take to prevent pipe failures is to:

- Specify designs in detail.
- Inspect thoroughly during and after construction to make sure that the design has been followed and that good engineering practice has been

followed when the design has not been specified in detail but left to the discretion of the construction team. Much more thorough inspection is needed than has been customary in the past.
- Perhaps try to explain to construction workers why it is important to follow the design in detail and to construct details not specified in the design in accordance with good engineering practice.

In the UK the Health and Safety Executive have put forward proposals for the inspection and certification of pressure systems[10] but they emphasize ongoing inspection throughout the life of the plant. Important as this is, the incidents described above suggest that inspection during and after construction will be much more effective in preventing failure.

Who should carry out inspections during and after construction? Construction teams usually employ their own inspectors, but they are often (or used to be) fairly junior people who do not carry much weight and often do not know what to look for. If radiographs are specified they see that they are done, but they may not have the knowledge to detect the departures from good engineering practice that are listed below. I suggest that inspections are carried out by:

- A member of the design team – he will notice departures from the design intention – and
- a member of the start-up team – he will have a greater incentive than anyone, as he will suffer from the results of the construction defects.

The following is a list of some of the things that should be looked for during these inspections. The numbers refer to the descriptions of the incidents.

During construction:
- Equipment has been made of the grade of steel specified and has received the specified heat treatment (A11, B2).
- Old pipe is not re-used without checking that it is suitable for the new duty (B11).
- Pipes are not laid underground (B4).
- Workmanship is of the quality specified and tests are carried out as specified (B3, B4, B15).
- Purchased equipment is undamaged (B8).

After construction and before start-up:
- Pipes are not secured too rigidly (B9).
- Pipes are free to expand and will not foul supports or other fixtures when they expand (A8, B10).
- Flanges on liquid lines are not located above cables (A8).
- Supports are correctly installed and assembled and springs are not fully compressed or extended (B6).
- There are no obviously substandard joints (A3).
- Pipes are not touching the ground (B12).
- Temporary supports have been removed (B1).
- Temporary branches, nipples and plugs have been removed and replaced by properly designed welded plugs (B16).
- Screwed joints have not been used (A9).

- Trace heating cannot be isolated on part of a line without isolating the whole (B14).
- Equipment has not been assembled wrongly. First identify any equipment which can be assembled wrongly (A2).
- Pipes do not pass through pits or depressions which can fill with water (A10).
- Relief valve tailpipes or drain lines are not so close to the ground that they may be blocked by ice or dirt (B17).
- Lines which contain water can be drained when not in use (C4).
- The slope of lines is correct, e.g. blowdown lines should slope towards the blowdown drum (B6).
- There is no bodging (B6, B7, B20).
- Reinforcement pads are vented (B18).
- There are no dead-ends in which water or corrosive materials can collect (B19). Note that dead-ends include little-used branches as well as blanked branches.
- There are no water traps in which water can collect (B19).
- Bellows are not distorted and any support rings are not loose (B8).

After start-up:
- Pipes are not vibrating (B13).

Note that the incidents do not suggest that there is anything wrong with the piping codes. (Item A12 described a rare incident due to an error in a code.) We do not need stronger, thicker pipes. The failures occurred because the codes were not followed or because the plant was not constructed in accordance with good engineering practice. This is a rather elusive quality that is not usually written down. An attempt to list a few of the factors that contribute towards it has been made above. The list is not intended to be complete. It merely summarizes the points that come out of the incidents described. It is intended as a starting point to which readers can add their own experiences. The construction inspector has to look out for things that no one has ever dreamt of specifically prohibiting.

Another review of pipe failures[11] is broadly in agreement with this chapter but attaches more importance to operating errors.

Finally, if we can reduce the quantity of hazardous material in the plant, or use a safer material instead, as discussed in Chapters 8 and 9, we need worry less about pipe failures.

References

This chapter is based on a paper which was published in *Plant/Operatons Progress*, Vol. 3, No. 1, Jan. 1984, p. 19, and thanks are due to the American Institute of Chemical Engineers for permission to quote from it.
1. DAVENPORT, J. A. *Chemical Engineering Progress*, Vol. 73, No. 9, Sept. 1977, p. 54.
2. KLETZ, T. A. *Myths of the Chemical Industry*, Institution of Chemical Engineers, Rugby, UK, 1985, p. 9.
3. VERVALIN, C. H. *Fire Protection Handbook for Hydrocarbon Processing Plants*, Vol. 1, 3rd edition, 1985, p. 122.
4. PRESCOTT, G. R. *et al.*, *Plant/Operations Progress*, Vol. 5, No. 3, July 1986, p. 155.
5. LLOYD, W. D. *Plant/Operations Progress*, Vol. 2, No. 2, April 1983, p. 120.

6. LAWRENCE, G. M. *Plant/Operations Progress,* Vol. 5, No. 3, July 1986, p. 175.

7. GEISLER, V. G. *Loss Prevention,* Vol 12, 1979, p. 9.

8. Safety Recommendations Nos. P-75-14 & 15, US National Transportation Safety Board, Washington DC, 14 Nov. 1975.

9. Health and Safety Executive, *Manufacturing and Service Industries 1982,* Her Majesty's Stationery Office, London, 1983, p. 15.

10. Health and Safety Executive, *Consultative Document: Proposed Pressure Systems and Transportable Gas Containers Regulations and Approved Codes of Practice,* Her Majesty's Stationery Office, London, 1985.

11. BLYTHING, K. W. and PARRY, S. T. *Pipework Failures – A Review of Historical Incidents,* UK Atomic Energy Authority, 1987.

Chapter 17

Conclusions

This chapter summarizes those recommendations that have appeared in many of the preceding chapters. Note that many important recommendations are not included, as they appear in only one or two of the preceding chapters. For example, errors in the preparation of equipment for maintenance, a major cause of accidents, appear only in Chapter 5. The accidents described were chosen primarily to illustrate the need for analysis in depth. Many accidents due to the most important immediate technical causes are described in reference 1.

Effective prevention lies far from the top event

In many of the accidents discussed, too much reliance was placed on measures which were designed to prevent the upper events in the diagrams (or the upper events in fault trees). If the measures failed, for any reason, there was then little or no opportunity for further defence.

The incident described in Chapter 4 is the most extreme example. Everyone was casual about leaks of ethylene because they had, they thought, eliminated all causes of ignition, and therefore the leaks could not ignite. When a source of ignition appeared, an explosion was inevitable. Those concerned were in the position of a military commander who neglects all the outer lines of defence because he considers the inner stronghold impregnable.

Bhopal (Chapter 10) is another example. There was extensive provision for dealing with high pressures in vessels containing methyl isocyanate (MIC): a cooling system to lower the temperature and thus the pressure, relief valves, a scrubbing system and a flare system. When the cooling, scrubbing and flare systems were not in full working order, a discharge to atmosphere was inevitable. A more effective loss prevention programme would have avoided an intermediate stock of MIC, kept the public away from the plant and tackled the underlying reasons that lay behind the failure to maintain the safety equipment – such as the lack of training and education.

It may be useful to summarize the main lines of defence that can be used to prevent major leaks of hazardous materials, starting with those that lie furthest from the top event. Some apply only to flammable materials.

146

(1) Avoid large inventories of hazardous materials by intensification, substitution or attenuation. (See below.)
(2) Inspect thoroughly during and after construction. (See Chapter 16.)
(3) Install gas detectors so that leaks are detected promptly. (See Chapter 4.) (Note: this does not remove the need for regular tours of inspection by operators. Even on plants which are fitted with gas detectors about half the leaks that occur are detected by people.)
(4) Warn people when a leak occurs. Those who are not required to deal with a leak should leave the area, by a safe route. (See Chapter 5.)
(5) Isolate the leak by means of remotely operated emergency isolation valves. (See Chapter 4.)
(6) Disperse the leak by open construction, supplemented, if necessary, by steam or water curtains. (See Chapter 4.)
(7) Remove known sources of ignition.
(8) Protect against the effects of the leak, as follows:
    Fire: Insulation and water spray.
    Explosion: Strengthened buildings,
                distance (i.e. prevent development nearby).
    Toxicity: Distance (i.e. avoid concentrations of people nearby).
(9) Provide fire-fighting and other emergency facilities.

This list, of course, takes no account of the software measures such as hazard and operability studies, audits etc., some of which are discussed below.

The control of plant modifications

Many of the accidents (see Chapters 1, 7, 8, and 14, and item 6 of Chapter 15) occurred because modifications of plant and process had unforeseen and undesirable side-effects, and the actions that should be taken to prevent similar accidents in the future were summarized in Chapter 7. (See also references 1 and 2 of Chapter 7.) People dealing with complex systems tend to think in straight lines. They think about the effects of their actions on the path to the immediate goal but remain unaware of the side-effects[2].

Testing and inspection of protective equipment

Many of the accidents would not have occurred if protective equipment, of various sorts, had been kept in working order. (See Chapters 1, 2, 3, 6, 7, 10 and 12.) To prevent similar accidents in the future, we need a two-pronged approach:

(A) An education programme to convince people, at all levels, that safety equipment can and should be kept in working order and that it is not an optional extra, something that can be neglected or put to one side under pressure of work. (See Chapter 6.) A one-off programme after an accident is not sufficient. An ongoing programme is necessary. An occasional lecture or piece of paper is not sufficient. It is better to involve people in regular discussions on accidents that have occurred, why they occurred and the action necessary to prevent them happening again (see Introduction), and to circulate regular reminders, in well-written and attractive publications, of the accidents that have happened and the precautions necessary. Compare your company's safety literature with that prepared to attract your customers!

Training in loss prevention is particularly important during a person's formative years as a student. Such training is standard practice in the UK but not in most other countries[3].

(B) An audit or inspection programme to make sure that equipment is being kept in working order. Much equipment should be tested at regular intervals: instruments monthly, relief valves every year or two, some equipment more often, e.g. the vents in Chapter 7. Managers should make spot-checks from time to time and there should be occasional audits by outsiders[4].

If protective equipment has to be disarmed, this should be signalled clearly so that everyone is aware of the fact, and is constantly reminded. (See Chapter 3.) In some cases it should be difficult or impossible to disarm the protective features. (See Chapter 12.)

User-friendly designs

'Friendly' is used to describe plants which will tolerate departures from ideal operations or maintenance without an accident occurring. Thus Bhopal would not have occurred if there had not been a large, unnecessary intermediate stock of MIC. The accident described in Chapter 6 would not have occurred if the recovered raw material had not been stored but fed straight back into the plant. 'What you don't have, can't leak (or explode)'. We should keep stocks of hazardous materials to the minimum (intensification) or use safer materials instead (substitution) or use the hazardous materials under the least hazardous conditions (attenuation)[5].

The plant described in Chapter 4 had numerous crossovers between parallel streams, providing many opportunities for leaks (and for errors). The closed compressor house magnified the effects of any leaks.

In the unit described in Chapter 2, the protective system was neglected and thus introduced a greater hazard than the one it was designed to remove, but the need for the system could have been avoided by moving the unit a few metres.

The nuclear reactors at Three Mile Island (Chapter 11) and Chernobyl (Chapter 12) were less friendly than gas-cooled reactors as they were dependent on added-on cooling systems which were liable to fail and gave the operators less time in which to respond. Chernobyl was particularly unfriendly, as it had a positive power coefficient (i.e. as it got hotter, the rate of heat production increased).

Carry out hazard and operability studies

Many of the accidents described show the need for critical examination of the design by hazard and operability studies (hazops) or similar techniques. (See Chapters 8–10.) The technique is now well-known[6,7] and widely used and there is therefore no need to describe it here. Less widely recognized is the need for similar (but shorter) studies in the earlier stages of design if we are to avoid some of the unfriendly features just described. Two such early studies are needed, one at the conceptual stage when we are deciding what process to use and where the plant is to be located, and one at the flowsheet stage[5]. A conceptual study at Bhopal, for example, might have queried the choice of product (other insecticides are available), the process to be used and the need for intermediate stocks. A flowsheet study would have allowed people to query the need for intermediate stocks again, the capacity of the scrubbing, flare and refrigeration systems and the need for sparage. The final hazop, on the line diagrams, would have allowed ways in which water might enter the MIC system to be explored.

Many companies will argue that they would have discussed all these subjects. This is true, but what is lacking in many companies is a structured, systematic, formalized technique in which every possible deviation is considered in turn for each line in the plant. Because modern designs are so complicated we cannot foresee the effects of deviations, or alternatives, unless we go through the design, bit by bit, slowly and systematically.

Samuel Coleridge described history as 'a lantern on the stern', illuminating the hazards the ship has passed through rather than those that lie ahead. It is better to illuminate the hazards after we have passed through them than not illuminate them at all, as we may pass the same way again, but it is better still to illuminate the hazards that lie ahead. This book is a lantern on the stern. Hazop is a lantern on the bow.

To use a different metaphor, like the chameleon we need to keep one eye on the past and one on the future[8].

Better management

Some necessary, but sometimes neglected, management features have already been discussed: audits, regular testing of protective equipment and hazard and operability studies.

In the incidents described in Chapters 1, 6, 7, 10, 11 and 12, operator training, or rather the lack of it, was a significant factor. Operators do not just need training in their duties but also need an understanding of the process and the hazards. Managerial ignorance was illustrated in Chapters 4, 5, 8 and 14.

A common feature has been failure to learn from the experience of the past. Sometimes the knowledge was forgotten, and sometimes it was not passed on to those who needed to know. Sometimes the whole company or factory failed to learn (see Chapters 4, 5 and 13), sometimes individual managers (see Chapters 2 and 7). The people involved in an accident do not forget but after a while they leave and others take their place. Organizations have no memory. Only people have memories and they leave.

There is much that can be done to learn from the experience of those who have been involved in accidents and to keep alive the memory of the past. For example, we can:

- Discuss accidents from time to time, as described in the Introduction.

- Describe old accidents as well as recent ones in safety bulletins and discuss them at safety meetings.
- Include, in standards and codes of practice, notes on the accidents that led to the recommendations.
- Keep a 'black book' in each control room – a folder of reports on accidents that have occurred. Do not include falls and bruises, but only accidents of technical interest, and include accidents from other plants. The black book should be compulsory reading for newcomers, at all levels, and old hands should dip into it from time to time to refresh their memories.
- Make more use of information storage and retrieval systems so that if we become interested in, say, explosions in compressor houses, it is easier to find reports on incidents that have occurred and the recommendations made.

As with the need to keep protective equipment in working order, we need more than an occasional one-off action; we need an ongoing programme.

Many of the accidents had a big impact on the public (see Chapters 8–12) and emphasize the need for hazardous industries to explain their hazards, and the precautions taken, rather better than they have done in the past if public disquiet is not to lead to official restrictions on their activities.

Underlying these various specific recommendations, the culture or climate of a company affects the way the staff behave, the action they take to prevent accidents and the enthusiasm with which they take it, as Chapter 3 (Report 5) makes clear. Different companies, even different plants within the same company, can have quite different climates. The climate depends to some extent on the training the staff have received, both in the company and during their formative years as students, as discussed in Chapter 10.

One of the finest US loss prevention engineers, the late Bill Doyle, used to say that for every complex problem there is at least one simple, plausible, wrong solution. This book confirms his opinion. It has shown that there is no 'quick fix' to the problem of preventing accidents. Each accident is different and complex; there is no single cure for the lot; each requires action at various levels – action to prevent the events that occurred immediately beforehand, action to remove the hazard and action to improve the management system. The first and third are usually possible on every plant but removing the hazard is often difficult on an existing plant. How far should we go? This is discussed in the next section.

How far should we go in bringing old plants and modern standards?

There is no simple answer to the question but a number of examples will show how the problem can be approached[9]. If we consider extreme examples first, the question is easy to answer. Some features cannot possibly be added to old plants, while others are as easy to install on an old plant as on a new one.

Standards of plant layout have improved in recent years and it is now normal to leave spaces between the sections of a large plant, or between small independent units, to restrict the spread of fire and to provide access for fire-fighting vehicles. Little or nothing can be done to improve the layout of existing plants, and if the layout is unsatisfactory all we can do, short of closing down the plant, is to compensate for the poor layout by higher standards of fire protection. Similarly, we now try to locate plants away from residential areas but housing has grown up around some old plants. Short of removing the houses or plants there is nothing that we can do.

Strengthened control rooms have been built on many new plants but it is almost impossible to replace an existing control room (though it has been done on one or two plants). However, we can remove large picture windows and heavy light fittings, and use special glass in any remaining windows (or glass protected by plastic film).

In contrast, gas detectors are as easy to install on an old plant as on a new one. Improved fire insulation is almost as easy to install on an old plant. Remotely operated emergency isolation valves are rather more difficult to install, as lines have to be cut and welded, but have been successfully fitted on many old plants.

The ventilation of many old compressor houses has been improved by knocking down some or all of the walls. They are not usually load-bearing. The most difficult part of the operation may be relocating equipment which is fixed to the walls.

Improvements to the drainage are more difficult. On many old plants the ground is sloped so that spillages run towards sources of ignition (see Chapter 3) or collect underneath the plant. It is often difficult or impossible to regrade the slope of the ground, but this has been done in a few cases. Some old plants (see Chapter 5) have surface drains. Installing underground drains is very expensive and disruptive and is rarely attempted.

A liquid-phase hydrocarbon processing plant, actually the scene of the accident described in Chapter 5, was constructed using compressed asbestos fibre (caf) gaskets. After a number of flange leaks had occurred, a change to spiral-wound gaskets was suggested but the size of the task – there were thousands of joints – appeared daunting. Further consideration led to a practicable solution. It was decided to replace all the gaskets in lines carrying liquid but not those in vapour lines. Three thousand gaskets were replaced over two years, during the normal six-monthly shutdowns, joints exposed to liquids above their boiling point being changed first. This example shows how an improvement to an existing plant which at first appears too large to contemplate becomes practicable when it is pruned a little and spread over several years.

A similar example is provided by item 10 of Chapter 15. After a screwed joint on an old plant had leaked, many, but not all, screwed joints were replaced. In general, screwed joints attached to main process lines were replaced but those beyond the first isolation valve, which could be isolated if they leaked, were left alone.

Some older plants contain mild steel pipes which can become too cold, and therefore brittle, under abnormal operating conditions, and this has caused some serious leaks and explosions[10]. Complete replacement of the mild steel lines is obviously impracticable. One company carried out a detailed study of its older pipework to determine its 'fitness for purpose'. Some lines were replaced; some were radiographed, to confirm that the welds were free from defects, and then stress-relieved, thus making them able to withstand lower temperatures. In other cases, additional low-temperature alarms or trips were installed, or the importance of watching the temperature was emphasized in operator training.

Replacement of relief valves which are known to be undersized is straightforward but enlargement of the flare header into which the valves discharge is more difficult. Trip systems have therefore been used to avoid or reduce the need for larger relief valves or to avoid a corresponding increase in the size of the flare header[11].

Although we cannot always bring the hardware on our old plants up to the latest standards ('inconsistency is the price of progress') we can make sure that the software (i.e. the methods of operation, training, instructions, testing and inspection, and auditing) are fully up to standard. This is at least half the problem. The earlier chapters have shown that if we want to prevent accidents, changes to the software are just as necessary as changes to the hardware. Safety by design is a desirable objective, and should always be our aim, but is not always possible, or practicable.

To sum up this section:

(1) If a change is as easy, or almost as easy, to make on an old plant as on a new one, then it should be made, unless the plant is due to be shut down in a few years or there is another way of achieving the same objective.
(2) We should not accept a higher standard of risk (to operators or the public) on an old plant than we would on a new one, but the methods we use to obtain that standard may be different. We may have to rely more on improvements to the software. If we do decide to do so, this must not be mere lip-service. We must be prepared to demonstrate that we have really improved the software.
(3) Newer plants often get the best managers and operators. Yet old plants, with their greater dependence on high standards of operation and management, may be in greater need of them. New plants have the glamour, and people like to be transferred to them, but we must take care that old plants are not starved of resources.
(4) When new standards are designed to prevent damage to plant or loss of production rather than injury to people, it is legitimate to leave old plants unchanged. Whether or not we do so should depend on the balance between probabilities and costs.
(5) We need to know how far our plants fell short of modern standards. Hazard and operability studies and relief and blowdown reviews can tell us.

References

1. KLETZ, T. A. *What Went Wrong?*, Gulf Publishing, Houston, Texas, 1985.
2. REASON, J. *Bulletin of the British Psychological Society*, Vol. 40, April 1987, p. 201.
3. KLETZ, T. A. *Plant/Operations Progress*, forthcoming.
4. KLETZ, T. A. *Health and Safety at Work*, Vol. 4, No. 3, Nov. 1981, p. 20.

5. KLETZ, T. A. *Cheaper, Safer Plants,* Institution of Chemical Engineers, Rugby, UK, 2nd edition, 1985.
6. KLETZ, T. A. *Hazop and Hazan,* Institution of Chemical Engineers, Rugby, UK, 2nd edition, 1986.
7. LEES, F. P. *Loss Prevention in the Process Industries,* Butterworths, 1980, Vol. 1, Section 8.9.
8. Malagasy proverb, quoted by A. Jolly, *National Geographic Magazine,* Vol. 171, No. 2, Feb. 1987, p. 183.
9. KLETZ, T. A. *Loss Prevention,* Vol. 14, 1981, p. 165.
10. VAN EIJNATTEN, A. L. M. *Chemical Engineering Progress,* Vol. 73, No. 9, Sept. 1977, p. 69.
11. KLETZ, T. A. *Myths of the Chemical Industry,* Institution of Chemical Engineers, Rugby, UK, 1984, p. 3.

Appendix: Some questions to ask during accident investigations

This questionnaire is intended to help up think of some of the less obvious ways of preventing an accident happening again. It will be more effective if the questions are answered by a team rather than a person working alone. The team leader should be reluctant to take 'Nothing' or 'We can't' as answers. He should reply, 'If you had to, how would you?'

What equipment failed?
 How can we prevent failure or make it less likely?
 How can we detect failure or approaching failure?
 How we control failure (i.e. minimize consequences)?
 What does this equipment do?
 What other equipment could we use instead?
 What could we do instead?

What material leaked (exploded, decomposed etc.)?
 How can we prevent a leak (explosion, decomposition etc.)?
 How can we detect a leak or approaching leak (etc.)?
 What does this material do?
 Do we need so much of it?
 What material could we use instead?
 What could we do instead?

Which people could have performed better? (Consider people who might supervise, train, inspect, check, or design better than they did as well as people who might construct, operate and maintain better than they did.)
 What could they have done better?
 How can we help them to perform better? (Consider training, instructions, inspections, audits etc. as well as changes to design.)

What is the purpose of the operation involved in the accident?
 Why do we do this?
 What could we do instead?
 How else could we do it?
 Who else could do it?
 When else could we do it?

Index

f indicates that the item extends over the following pages